T0258082

IEE TELECOMMUNICATIONS SERIES 29

Series Editors: Professor J. E. Flood
Professor C. J. Hughes
Professor J. D. Parsons

ATM

THE BROADBAND TELECOMMUNICATIONS SOLUTION

Other volumes in this series:

ATM

THE BROADBAND
TELECOMMUNICATIONS
SOLUTION

L G Cuthbert
and
J-C Sapanel

The Institution of Electrical Engineers

Published by: The Institution of Electrical Engineers, London,
United Kingdom

© 1993: The Institution of Electrical Engineers
Reprinted 1998

The Institution of Electrical Engineers,
Michael Faraday House,
Six Hills Way, Stevenage,
Herts. SG1 2AY, United Kingdom

British Library Cataloguing in Publication Data

A CIP catalogue record for this book
is available from the British Library

ISBN 0 85296 815 9

Printed in England by Short Run Press Ltd., Exeter

Contents

Preface

In the last few years there has been a rapidly increasing interest in having broadband *wide-area* communications. Most people working in a technical environment are well-used to having moderately fast communications through their local area networks. The possibility of adding the capability of video and multi-media support, together with extending the connectivity world-wide, has generated considerable excitement, and work, in the telecommunications industry. The pace of these developments can be gauged by the fact that ATM was adopted as the transfer mode for broadband ISDN in the beginning of 1988, the choice of cell-format was made in 1989, yet now in 1993 we already have the first experimental networks with real pilot applications being planned and implemented.

This book sets out to explain the principles of ATM, the issues involved in ATM networks and the reasons that applications based on ATM are generating so much interest. It is aimed at a wide audience: those who work in telecommunications or are studying it and who wish to gain an understanding of the techniques that will change the face of telecommunications across the world.

The material in the book comes from the work of the group studying network planning in the RACE project 'R1022: Telecommunications for ATD'. Although the book has been edited by LC and JCS, the material within it has been produced jointly by all members of the group:

João Bastos, CET, Telecom Portugal
Denis Carette, BELGACOM, Belgium
Laurie Cuthbert, Queen Mary and Westfield College, UK
Neil Henderson, BT Laboratories, UK
Beat Keller, Ascom Tech, Switzerland
Tassos Lyratzis, National Technical University of Athens, Greece
Jonathan Pitts, Queen Mary and Westfield College, UK
Jean Claude Sapanel, CNET, France Telecom

May 1993

Acknowledgements

This book stems from the work done by the project 'R1022: Technology for ATD' as part of the RACE programme of the Commission of the European Communities. The authors would like to acknowledge the financial support of the Commission under that programme.

Work within R1022 was divided into 'taskgroups' and that which led to this book was the responsibility of taskgroup V: 'Introduction Strategies'. This group studied the general network issues involved in introducing B-ISDN, the various options in an evolution from narrowband to broadband and also performed some techno-economic studies. During the period over which the group worked, there were several changes in membership and the authors, the members of the group during 1992, would like to express their thanks for all the contributions made by the following earlier members:

- Miltos Anagnostou, National Technical University of Athens, Greece
- Olivier Beau, Alcatel CIT, France
- Wes Carter, BT Laboratories, UK (who led the group from the start of the project in 1988 until 1990)
- Emmanuel Desmet, Alcatel Bell, Belgium
- Loic Etesse, CNET, France Telecom
- José Maio, CET, Portugal
- Phillip Ogunbono, BNR Europe (the STC Technology Ltd.), UK
- Sathya Rao, Ascom Tech, Switzerland
- Eric Usher, BNR Europe (the STC Technology Ltd.), UK
- Albrecht Widmaier, Alcatel SEL, Germany

Other people have also contributed in many ways, particularly in reviewing the project deliverables on which it was based, and we would like to say thank-you in particular to the following for those comments:

- Paul Verbeek, Alcatel Bell, Belgium (the project manager of R1022)
- Peter Heuer, FIDBP, Germany
- Günther Kettler, FIDBP, Germany

Finally, we would like to record our appreciation for the work of the group (taskgroup III) which worked within the project in the area of traffic engineering whose work has been incorporated in the chapters on 'Traffic Control' and 'Traffic Engineering'. This group consisted of a large number of people from a wide range of organisations (universities, network operators and industry) across Europe. It would be impracticable to list here all the members of taskgroup III so we would like to express our thanks by acknowledging the leaders on behalf of the whole group:

- Paul Kühn, University of Stuttgart, Germany
- Ralf Lehnert, PKI, Germany.

Abbreviations

AAL	ATM Adaptation Layer
AAL-PCI	AAL Protocol Control Information
AAL-SDU	AAL Service Data Unit
AAL-IDU	AAL Interface Data Unit
ACX	ATM Cross-connect
AF	Access Facility (within MANs)
APON	ATM Passive Optical Network
ATD	Asynchronous Time Division
ATM	Asynchronous Transfer Mode
ATM-SDU	ATM (layer) Service Data Unit
AUU	ATM-layer-user-to-ATM-layer-user parameter
B-ISDN	Broadband ISDN
B-NT2	Broadband NT2
B-NT1	Broadband NT1
BER	Bit-Error Rate or Ratio (context distinguishes)
BPON	Broadband Passive Optical Network
C-4	SDH C-4
CAC	Connection Admission Control
CAD	Computer Aided Design
CATV	Closed Area TV (Common Antenna TV)
CBDS	Connectionless Broadband Data Services
CBR	Constant Bit-Rate (service)
CC	Cross-Connect or Country Code (context distinguishes)
CEQ	Customer EQuipment
CFS	(RACE) Common Functional Specification
CLP	Cell-Loss Probability
CLR	Cell-Loss Rate or Ratio (context distinguishes)
CN	Customer Network (used in MAN context)
CPCS	Common Part Convergence Sublayer (in AAL)
CPCS-SDU	CPCS Service Data Unit
CPN	Customer Premises Network
CRC	Cyclic Redundancy Check
CRF	Connection Related Functions
CS	Convergence Sublayer
CS-PDU	CS Protocol Data Unit

CSDN	Circuit-Switched Data Network
DBS	Direct Broadcast Satellite
DLCI	Digital Local Connection Identifier
DDI	Direct Dialling In
DECT	Digital European Cordless Telephone (standard)
DMPDU	Derived MAC Protocol Data Unit (PDU within MAN)
DQDB	Distributed-Queue, Dual-Bus (MAN standard)
ECU	European Currency Unit
FDDI	Fibre Distributed Data Interface
FMBS	Frame Mode Bearer Service
GFC	Generic Flow Control (field in ATM header)
GOS	Grade Of Service
GSM	Global System for Mobile communications
HEC	Header Error Control
IDU	Interface Data Unit
IMAI	Inter MAN ATM Interface
IN	Intelligent Network
ISDN	Integrated Services Digital network
ISUP	Integrated Services User Part (of No 7 signalling system)
IWU	InterWorking Unit
LAN	Local Area Network
LEX	Local EXchange
MAN	Metropolitan Area Network
MBS	Mobile Broadband Service
MSS	Man Switching System
N-ISDN	Narrowband ISDN
NNI	Network Node Interface
NPC	Network Parameter Control
NSC	Network Specialised Centre
NT1	Network Termination type 1
NT2	Network Termination type 2
OAM	Operations And Maintenance
OLR	Overall Loudness Rating
OSI	Open Systems Interconnection
PA	Pre-Allocated (in DQDB MAN)
PBX	Private Branch Exchange
PCI	Protocol Control Information
PDH	Plesiochronous Digital Hierarchy
PDU	Protocol Data Unit
PL-OAM	Physical-Layer OAM
PLMN	Public Land Mobile Network
PM	Physical Medium sublayer
PON	Passive Optical Networks
PSDN	Packet-Switched Data Network
PSTN	Public Switched Telephone Network

PT	Payload Type field (in ATM header)
PTO	Public Telecommunications Operator
QA	Queue Arbitrated slots (in DQDB MAN)
QOS	Quality Of Service
RU	Remote Unit
SAP	Service Access Point
SAR	Segmentation AND Reassembly (layer)
SAR-SDU	Segmentation AND Reassembly (layer) Service Data Unit
SAR-PDU	Segmentation AND Reassembly (layer) Protocol Data Unit
SCP	Service Control Point (in IN)
SDH	Synchronous Digital Hierarchy
SDU	Service Data Unit
SN	Sequence Number (in AAL)
SSCS	Service Specific Convergence Sublayer
SSP	Service Switching Point (in IN)
STM	Synchronous Transfer Mode
STM-1	1st level of SDH
STM-4	2nd level of SDH
TA	Terminal Adapter
TC	Transmission Convergence
TDM	Time Division Multiplex
TDMA	Time Division Multiple Access
TE1	Terminal Equipment type 1 (ISDN compatible)
TE2	Terminal Equipment type 2 (non-ISDN - needs TA)
TEX	Trunk (or Transit) EXchange
TMN	Telecommunications Management Network
TPON	Telephony on Passive Optical Network (used generically in this book)
TV	Television
UMTS	Universal Mobile Telephone Service
UNI	User-Network Interface
UPC	Usage Parameter Control
URS	Universal Roaming Subscribers
VBR	Variable Bit-Rate (service)
VC	Virtual Channel (in ATM context), Virtual Container (in SDH context)
VC-4	Virtual Container type 4
VCC	Virtual Channel Connection
VCI	Virtual Channel Identifier
VP	Virtual Path
VPC	Virtual Path Connection
VPI	Virtual Path Identifier

Chapter 1

Introduction

1.1. What is broadband?

Communication systems have developed rapidly over the last decade with new developments such as:

- Replacement of electromechanical switches with digital ones
- Introduction of optical transmission systems
- Migration of digital communication towards the customer with ISDN
- Rapid growth in mobile telephony
- High-speed data networks such as LANs

Why then is there a need for new, broadband systems? What actually is broadband?

Some would argue that moving towards broadband is unnecessary, that existing networks are now so well developed that they can carry all the traffic and services that a customer is likely to want. Others point to the parallel with data networks where there has been a revolution in the speeds being used: 64 kbit/s X.25 [43] networks became too slow for local area networks (LANs) and so were replaced with higher-speed systems such as Ethernet. These in their turn are becoming too slow to carry the volume of traffic required over a LAN for modern applications and so there is a move towards even higher-speed techniques, such as *fibre distributed data interchange* (FDDI) and *distributed-queue dual-bus* (DQDB).

The reality is that it is the *service* the customer pays for and this service is used to support an *application*. These ideas are explored more fully in Chapter 3. If new applications require new services running at a higher bit-rate, then the underlying communications network must evolve to support these new services.

The boundary between broadband and narrowband communications systems is commonly taken as being at 2Mbit/s. It is clear that in the world of data networks we already have broadband communications since LANs commonly operate at a speed of 10Mbit/s and an individual user can make use of well over 2Mbit/s at a particular time.

However, this broadband capability does not yet extend beyond the LAN into the Wide Area Network (WAN); nor is it integrated, since the customer can transmit data at over 2Mbit/s yet needs a separate 64 kbit/s circuit for telephone conversations.

It is the concept of providing an integrated broadband communications network (the Broadband Integrated Services Digital Network: the B-ISDN) which led to the development of asynchronous transfer mode (ATM) as the transfer mode capable of supporting all of these new services and its adoption by the CCITT as the target transfer mode for the B-ISDN.

A great deal of work is currently taking place throughout the world on developing the B-ISDN, in terms of the enabling technologies and the underlying theoretical and experimental work to enable engineers to plan and dimension the new networks. Work is also taking place on the likely services to be carried, since it cannot be stated too often that the purpose of the network is to provide the services that the customer is willing to pay for. At the time of writing this book, the beginning of 1993, experimental ATM switches exist and manufacturers are announcing commercial systems; pilot ATM networks are on the point of being installed and it is expected that by the second half of the decade there will be a significant number of networks in use, many of them as private customer networks.

This book sets out to describe and explain the main features of a broadband ATM network, *from the network point of view*. It is not a book on technology, so it does not describe in detail the various techniques for switching ATM cells, particularly as many of the switching architectures are proprietary. ATM switching is described in References 55 and 66.

1.2. Why ATM ?

The term *ATM* stands for *asynchronous transfer mode*; it is a system whereby information is transferred asynchronously with its appearance at the input of communications system. Information is buffered as it arrives and is inserted into an ATM cell when there is enough to fill the cell, the cell is then transported across the network. At a multiplexing stage a cell from a particular stream is transmitted as soon as there is an unused ATM cell available to carry it; if there is no information to be transmitted an *unassigned cell* is transmitted instead (Figure 1.1). It is clear that the principle is very similar to that of a packet-switched network. However, ATM is different in several ways:

- The cells are much shorter than in a conventional packet network in order to achieve reasonable values for delay variance.
- Overhead is minimised in order to maximise efficiency at the high bit-rates used (for instance there is no error correction mechanism, unlike current X.25 packet networks [43]).
- Cells are transported at regular intervals; there is no space between cells, idle periods on the link carry unassigned cells.

- The order in which cells arrive is guaranteed to be the same as the order in which they were transmitted (*ATM is said to provide cell sequence integrity*). This may or may not be the case in other packet networks.

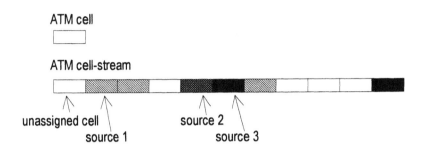

Figure 1.1 Basic principle of ATM cell-stream

Such a technique provides great flexibility, because it can match the rate at which it transmits cells to the rate at which the information is generated. This is important for many of the new high-bit-rate services that are being developed, particularly those with a video component, because they are *variable bit-rate* (VBR) services: the bit-rate for the source is not constant. Why this might happen is intuitively obvious if one thinks of a video picture where the scene changes from a commentator to a race scene, for instance, and where some form of compression technique is being used to reduce the amount of information that is actually transmitted. Most forms of compression rely on transmitting basically the *changes* in information: with the commentator there is only a low rate of information to be transmitted since the only changes are probably the facial changes associated with speaking; as the scene switches to the race itself there is a very large burst of information (because of the complete picture change) and then it settles to another rate; higher than that associated with the commentator because of the greater changing scene associated with the race.

ATM matches the transmission of the cells to the generation of information so it does not waste capacity during the low-activity periods of the source by transmitting wasted cells: the cells that could be transmitted during this period are available for other sources and so ATM provides inherent *statistical multiplexing* on the link. If there is a reasonably large number of VBR sources generating information to be transported over a link, then a significant statistical multiplexing gain can be achieved. This is illustrated (with a stylised representation of the VBR source requirements) in Figure 1.2.

Moreover, the inherent multiplexing of the cells with ATM offers ease of integration of sources onto one link so that the network operator only has to provide one connection (one *access link*) to the customer and all the services can be provided over this link. However, there are, as might be expected, disadvantages with ATM, the most significant two being *cell delay variation* and *cell assembly delay*.

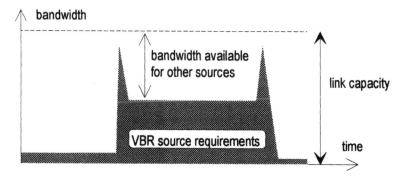

Figure 1.2 Principle of statistical multiplexing with VBR sources

Cell delay variation (CDV) arises from the variable delays introduced in the network by the queues at switches and multiplexers and this leads to a change from what would be expected in the gap between different cells. This is illustrated for a constant bit-rate (CBR) source in Figure 1.3. The problem of CDV is most acute with services where the difference in delay affects the quality of service perceived by the user, the so-called *delay-sensitive* services. A particular example of such a service is speech transmission, which is obviously an important service in B-ISDN.

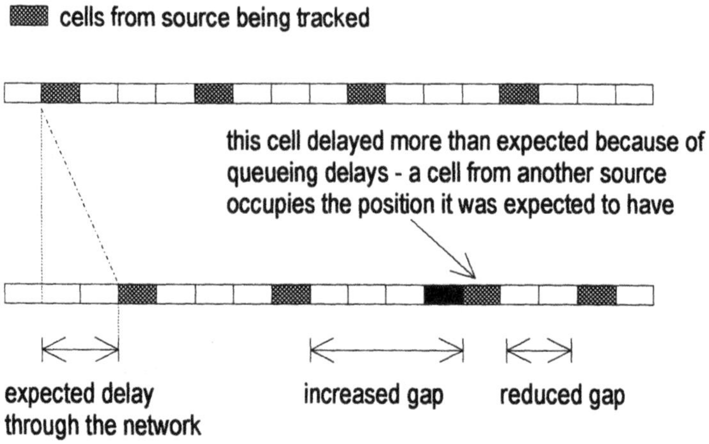

▓▓ cells from source being tracked

this cell delayed more than expected because of
queueing delays - a cell from another source
occupies the position it was expected to have

expected delay   increased gap   reduced gap
through the network

Figure 1.3 Principle of cell delay variation

The second problem, cell assembly delay, arises because information from a source is buffered until there is sufficient to fill a cell: this is 48 octets of information. The time the information waits in the buffer obviously depends upon the rate at which it is arriving and will be longer for low bit-rate sources.

Because the time spent in the buffer represents a delay it will have greatest impact on delay-sensitive services.

Voice telephony is again the most common service to be affected. At 64 kbit/s it takes 6ms to assemble 48 octets and this delay is significant when it comes to considering effects such as echo; this is considered in more detail in Chapter 5.

These two problems do not arise with the synchronous transfer mode used in the conventional narrowband ISDN (N-ISDN) and the question has been put as to whether the flexibility and statistical multiplexing achievable with ATM is worth the extra complications in delay-sensitive services, particularly in voice telephony. The answer to this question is *yes* because the international standards body (the CCITT) has agreed (in 1988) that ATM will be the target transfer mode for B-ISDN. The problems in delay-sensitive services are seen as providing a challenge to telecommunications engineers rather than being an insurmountable obstacle to the introduction of ATM.

1.3. Transfer mode

The previous section talked about ATM as a *transfer mode*. Information is transferred across the network in cells asynchronously from the actual arrival of the information at the input to the network. However, it is most important to appreciate that this transfer is independent of the actual transmission system used: ATM cells can be transmitted within the frames of the synchronous digital hierarchy (SDH) or in the frame structure of the current plesiochronous hierarchy: in each case there is a different mapping between the ATM cells and the bits of the transmission system.

1.4. CCITT standardisation of B-ISDN

ITU (International Telecommunication Union) is the institution of the United Nations Organisation dealing with telecommunications. CCITT (Comité Consultatif International Téléphonique et Télégraphique) and CCIR (Comité Consultatif International des Radiocommunications) are the standardisation bodies. At the time of writing this book, the name of CCITT was being changed to ITU-TSS (Telecommunications Standards Sector) but for familiarity the references to CCITT standards have retained the old form.

CCITT works in study period of 4 years, the last ones being 1985-1988 and 1989-1992. At the beginning of a period a list of items called *questions* is established with the objective to produce recommendations at the end of the period. Other procedures which allow publication of recommendations during a study period also exist (accelerated procedure).

The 1985-1989 study period started the study of B-ISDN. Only one recommendation was adopted (Rec. I.121 - Broadband aspects of B-ISDN) which contained important agreement:

- ATM (asynchronous transfer mode) was agreed as the 'target transfer mode solution for implementing a B-ISDN'.
- The first draft of B-ISDN protocol reference model (layers and functions) was agreed.
- Only two bit-rates were agreed for the User Network Interface (UNI): 150 Mbit/s and 600 Mbit/s.

This 1989 recommendation contains many chapters that gave rise to recommendations in the following study period. Important issues such as the cell size were not determined but some guidance was given, e.g. header size in the range of 3 to 8 octets and information field size in the range of 32 to 120 octets.

At the beginning of the 1989-1992 study period, a consensus existed between the CCITT participants to standardise B-ISDN and to have a set of recommendations before the end of the period. At the first meeting in June 1989 the following decisions concerning the cell size and interfaces were agreed:

- The header size (5 octets) and the information field size (48 octets) were agreed.
- ATM is independent of transmission systems: cell multiplexing is done by the cell delineation mechanism.
- There should be two physical layers at the T interface (SDH-based and cell-based).

A list of recommendations was established during this meeting and the results of the CCITT work is contained in the following set of 13 recommendations which were agreed in 1990 (I.371 was separated out in June 1991). These are described in more detail in the Appendix.

I.113: *Vocabulary of terms for broadband aspects of ISDN*
I.121: *Broadband aspects of ISDN*
I.150: *B-ISDN functional characteristics*
I.211: *B-ISDN service aspects*
I.311: *B-ISDN general network aspects*
I.321: *B-ISDN protocol reference model and its application*
I.327: *B-ISDN functional architecture*
I.361: *B-ISDN ATM-layer specification*
I.362: *B-ISDN ATM adaptation layer (AAL) functional description*
I.363: *B-ISDN ATM adaptation layer (AAL) specification*
I.371: *Traffic control and resource management in B-ISDN*
I.413: *B-ISDN user-network interface*
I.432: *B-ISDN user-network interface; physical-layer specification*
I.610: *OAM principles of the B-ISDN access*

Chapter 2

Asynchronous Transfer Mode (ATM)

2.1. Basic principles of ATM

As explained in the previous chapter, ATM has been chosen as the target transfer mode for B-ISDN, since it offers a flexible transfer capability common to all services because of its independence of the bit-rate and data structure of the services carried. For these reasons, additional functionalities are needed to accommodate the various services and these are added at the edge of the ATM network.

ATM is a particular packet-oriented transfer mode that uses asynchronous time division multiplexing techniques with the multiplexed information flow being organised into blocks of fixed size, called *cells*. A cell (Figure 2.1) consists of an *information field* carrying user information (*cell payload*) and a *header* containing network information, for example routeing information. Because cells from more than one connection are multiplexed together, there has to be a means to identify cells belonging to the same connection and this is done by the information in the header.

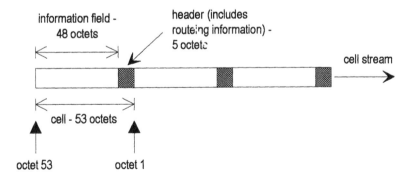

Figure 2.1 Basic ATM cell format

The information field is transported transparently by the ATM network and no processing (for example error control, used in conventional packet networks) is performed on this field by the ATM network. The cell sequence is preserved by the ATM network, cells being received in the same order as they are sent (*cell sequence integrity*).

Because ATM is a connection-oriented technique, a path has to be established between the users before information can be exchanged. This is done by the connection set-up procedure at the start and by a clear-down procedure at the end. The set-up procedure uses a signalling protocol for on-demand connection and other means, for instance a network-management procedure, for semi-permanent or for permanent connections. A broadband *call* can be a multi-media call having a number of components. As each component generally requires a separate *connection*, it is usual to discuss what happens at a 'connection' level rather than 'call' level.

Each connection has a *transfer capacity* (a bandwidth) assigned to it according to the user's request, subject to there being sufficient capacity available. This is usually done during the connection set-up procedure using a process called *connection admission control* (CAC); this process determines the parameters that the connection will be allowed to have depending on the user's needs. There is another process, *usage parameter control* (UPC) that monitors the connection and takes action if the connection attempts to exceed the limits that have been allocated to it.

Broadband ISDN follows the same principles as narrowband (64 kbit/s) ISDN whereby the user information and signalling information are carried on separate channels, the B and D channels, respectively.

2.2. Protocol reference model

In a similar way to the familiar OSI 7-layer model, the B-ISDN also has a protocol reference model (shown in Figure 2.2), which consists of a *user plane*, a *control plane* and a *management plane*. The concepts of *service access points* (SAPs), *service data units* (SDUs) and *protocol data units* (PDUs) that are found in the OSI layered approach [65] also apply to this protocol reference model.

The user plane (for user information transfer) and the control plane (call control and connection control functions) are structured in layers. Above the *physical layer*, the *ATM layer* provides cell transfer for all services and the *ATM adaptation layer* (always referred to as the *AAL*) provides service-dependent functions to the layer above the AAL. The management plane provides network supervision functions. As would be expected with such a layered protocol, the characteristics of the ATM layer are independent of the physical medium.

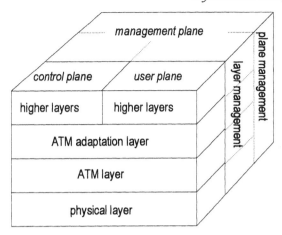

Figure 2.2 B-ISDN protocol reference model and the functions of the layer

The AAL and physical layer are further divided into sublayers and the functions performed within these are also shown in Table 2.1. A detailed explanation of the functions of the layers and sublayers is given later in this chapter.

Table 2.1 ATM protocol model functions

layer / sublayer	function
ATM adaptation layer	
convergence sublayer	convergence
segmentation & reassembly sublayer	segmentation & reassembly
ATM layer	generic flow control
	cell header generation/extraction
	cell VPI/VCI translation
	cell multiplex and demultiplex
Physical layer	
transmission convergence sublayer	cell-rate decoupling
	HEC header generation/verification
	cell delineation
	transmission frame adaptation
	transmission frame generation & recovery
physical medium sublayer	bit timing
	physical medium

2.3. ATM layer

It is worth considering the ATM layer first as this is the layer that is concerned with transporting information across the network.

ATM uses virtual connections for information transport and these connections are divided into two levels: the *virtual path* level and the *virtual channel* level. This subdivision of the transport function is one of the powerful features of ATM, as will be seen later.

The functions of the ATM layer are shown in Figure 2.2 and are considered in more detail below. All these functions are supported by the ATM cell header which is described later in this section.

Cell multiplexing and demultiplexing

In the transmit direction, the cell multiplexing function combines cells from individual *virtual paths* (VPs) and *virtual channels* (VCs) into one cell flow. In the receive direction, the cell demultiplexing function directs individual cells to the appropriate VP or VC.

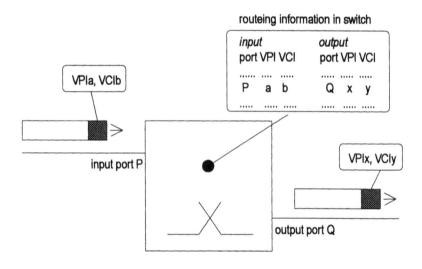

Figure 2.3 Illustration of the local properties of VPI and VCI value

Cell virtual path identifier (VPI) and virtual channel identifier (VCI) translation

The VPIs and VCIs are labels that identify a particular VP and VC on that link. The switching node uses the these values to identify a particular connection and will then, using the routeing information established at connection set-up, route the cells to the appropriate output port. The switching element changes the value of the VPI and VCI fields to new values that are used on the output link (Figure 2.3). There is a complication in that a *VP switch* will only change *VPI* values and will leave the VCI unchanged; this is discussed in detail in section 2.4.

Cell header generation/extraction

These functions apply at points where the ATM layer is terminated. In the transmit direction, the cell-header generation function receives a cell-information field from a higher layer and generates an appropriate ATM cell header, except for the HEC (*header error control*) sequence which is calculated and inserted by the physical layer. In the receive direction, the cell header extraction function removes the ATM cell header and passes the cell information field to the ATM adaptation layer.

Generic flow control

The generic flow control (GFC) function is used only at the *user-network interface* (UNI). It may assist the customer network in the control of cell flows *towards* the network but it does not perform flow control of traffic *from* the network. The GFC protocol information is not carried through the network.

The generic flow control could be used within a user's premises to share ATM capacity among terminals and some ideas about using this concept in ATM LANs have already been put forward.

2.3.1.   Cell structure

The ATM cell consists of a 5-octet header and a 48-octet information field immediately following the header; it is fully specified in CCITT Recommendation I.361 (B-ISDN ATM-layer specification). Two cell formats have been specified, one for the user-network interface (UNI) and the other for the network-node interface (NNI). The UNI format is used between the user installation and the first ATM exchange as well as within the user's own network. The NNI format is used between the ATM exchanges in the trunk network.

The header is divided into the following fields (Figure 2.4):

- *Generic flow control (GFC) field*: The GFC field contains 4 bits at the UNI and, as explained above, it can provide flow control information towards the network. At the time of writing, the exact mechanism has not been standardised.
- *Routeing field (VPI/VCI)*: 24 bits (8 for VPI and 16 for VCI) are available for routeing at the UNI and 28 bits (12 for VPI and 16 for VCI) at the NNI.
- *Payload type (PT) field*: 3 bits are available for payload type identification. The payload type field is used to provide an indication of whether the cell payload contains user information or network information; this is specified in CCITT Recommendation I.361. When the PT field indicates that the cell contains network information, the network processes the information field of the cell.

- *Cell loss priority (CLP) field*: This bit may be set by the user or service provider to indicate lower-priority cells. Cells with the CLP bit set are at risk of being discarded, depending on the conditions in the network.
- *Header error control (HEC) field*: the HEC field consists of 8 bits. This field is processed by the physical layer to detect errors in the header, the error control covering the entire cell header. The code used for this function is capable of either single-bit error-correction, or multiple-bit error-detection.

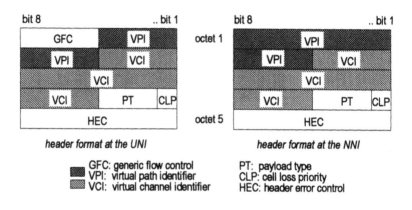

Figure 2.4 Format of the ATM header

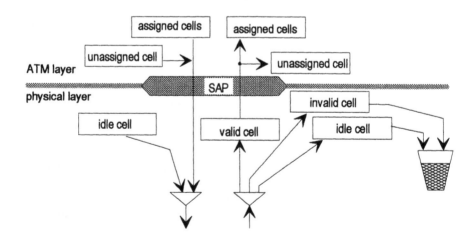

Figure 2.5 Insertion and extraction of types of ATM cells

2.3.2. Types of cell

The different types of cell are listed in Table 2.2. The type of cell containing user information is the *assigned cell*. It is important to note that the *idle cell* and the *invalid cell* only exist in the physical layer, as is shown in Figure 2.5. It is also worthwhile emphasising the difference between the *unassigned cell* and the *idle cell*. The unassigned cell is an unused cell in the ATM layer, that is a cell transmitted when there is no information to be carried; however, the idle cell is inserted by the physical layer in order to match the transmission rate to the transmission system, or for other physical-layer purposes.

Table 2.2 Different types of cell in the ATM layer

idle cell:	inserted/extracted by the physical layer in order to adapt the cell flow rate at the boundary between the ATM layer and the physical layer to the available payload capacity of the transmission system
valid cell:	has a header with no error or which has been corrected by the HEC verification process
invalid cell:	a cell with a header that has errors that have not been modified by the HEC verification process (discarded at physical layer)
assigned cell:	provides a service to an application using the ATM-layer service
unassigned cell:	an ATM-layer cell which is not an assigned cell

2.4. ATM objects

The generic term *ATM object* is used to cover the four types of element that can be established at the ATM-layer:

- *Virtual channel link* - identified by a *virtual channel identifier* (VCI)
- *Virtual channel connection* - a concatenation of virtual channel links
- *Virtual path link* - identified by a *virtual path identifier* (VPI)
- *Virtual path connection* - a concatenation of virtual path links

At a given interface, in a given direction, there are different *virtual path links*, each of which has its own VPI and is multiplexed onto the same physical-layer connection. Within a virtual path there are *virtual channel links*, each of which has its own VCI. This is shown diagrammatically in Figure 2.6.

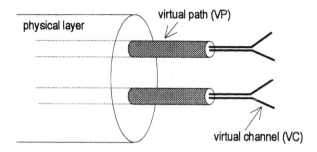

each VP within the physical layer has a different VPI value
and each VC within a VP has a different VCI value

Figure 2.6 Conceptual view of virtual channels and virtual paths within the physical layer

2.4.1.   Virtual channels

The basic concept in the establishment of an end-to-end connection is the request for a series of links from source to destination, the decision on whether to grant the request for the use of a link being dependent on the resources requested and the capacity remaining on the link. The series of *virtual channel links* thus established is called a *virtual channel connection* (VCC). The virtual channel is identified in each cell by a virtual channel identifier (VCI) which forms part of the cell header. Routeing translation tables in each VC switching node provide VCI translation information for every cell going into the switch. The relevant information for these translation tables is entered during call establishment and remains constant for the duration of a call. Within a virtual channel link the VCI has a particular value, but it will change from link to link within a virtual channel connection (Figure 2.7).

Virtual channel connections can be used in a variety of ways:

- *User-to-user applications*: In this type of application, the VCC is between the customer's equipment at each end of the connection, i.e. it extends to the S or T reference point at each end. The information is carried in ATM cells from one customer's equipment to the other's.
- *User-to-network application*: In this application, the VCC is between a customer's equipment and a network node and provides access to a network element.
- *Network-to-network application*: In this application, the VCC extends between two network nodes. The network-to-network application of this VCC includes network traffic management and routeing.

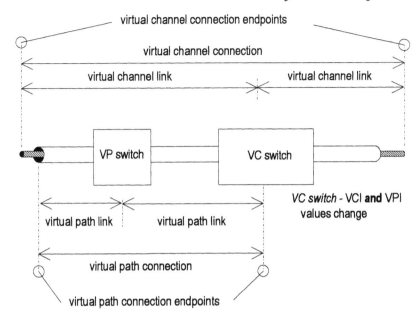

Figure 2.7 Illustration of the difference between VC (and VP) links and connections

Virtual channel connections have the following general properties:

- A user of a VCC is provided with a quality of service specified by parameters such as cell-loss ratio and cell delay variation.
- VCCs can be provided on a switched or (semi-)permanent basis.
- Cell sequence integrity is preserved within a VCC - a fundamental principle of ATM.
- Traffic parameters are *negotiated* between a user and a network for each VCC when the connection is set up and may be subsequently re-negotiated. Input cells from the user to the network are monitored (by the *usage parameter control* function) to ensure that the negotiated traffic parameters are not violated. If they are violated, the UPC function may take action against cells from that connection, such as discarding the offending cells.

The setting-up and release of VCCs at the user-network interface can be performed in various ways:

- Without using signalling procedures, e.g. by subscription, (permanent or semi-permanent connections)
- Meta-signalling procedures, whereby a special VCC (a meta-signalling VCC) is used to establish or release a VCC used for signalling. This is a new concept compared to narrowband ISDN, where the D channel used for signalling is always present on the user access. In B-ISDN, it is possible to have several

signalling channels. A protocol called the meta-signalling protocol is used to establish and release these signalling channels. The protocol is carried in a standardised channel for which a VCI value is allocated.

- User-to-network signalling procedures, e.g. using a signalling VCC to establish or release a VCC used for end-to-end communications
- User-to-user signalling procedures, e.g. using a signalling VCC to establish or release a VCC within a pre-established VPC between two UNIs.

The value assigned to the VCI during the call set-up described above could be assigned by the network, by the user's equipment, by negotiation between the user's equipment and the network or even pre-set by standardisation.

The specific value assigned to a VCI at a UNI is, in general, independent of the service provided over that VC. However, in order to make it easier to interchange terminals, and to perform initialisation on terminals, it is desirable to use the same value (a *pre-assigned VCI*) for certain functions on every UNI. For example, the same VCI value for the meta-signalling VC will be used on all UNIs in order to simplify initialisation of the terminal equipment.

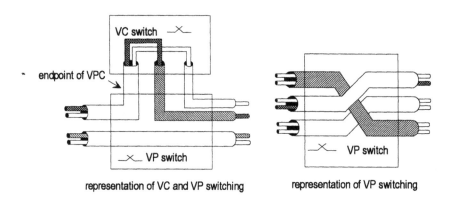

representation of VC and VP switching   representation of VP switching

Figure 2.8 Representation of VC and VP switching

2.4.2. Virtual paths

A *virtual path* (VP) is a generic term for a bundle of virtual channel links, all the VC links in the bundle having the same endpoints.

Virtual path links are concatenated to form a *virtual path connection* (VPC) that extends between two VPC endpoints or, in the case of point-to-multipoint arrangements, between more than two VPC endpoints. A VPC endpoint is the point where the VCIs are originated, translated or terminated.

The concept of the virtual path is to provide logical direct routes (shared by several VCs) between switching nodes via intermediate cross-connect nodes. A virtual path is an arrangement (semi-permanent or switched) that provides the logical equivalent of a link between two switching nodes that are not necessarily directly connected by a single physical link. Thus, it allows a distinction between physical and logical network structure and gives flexibility to rearrange the logical structure according to the traffic requirements. (See Figure 2.8.)

A virtual path is identified in a cell by a *virtual path identifier* (VPI) that forms part of the cell header. Routeing translation tables in each cross-connect node provide VPI translation for each cell entering a cross-connect. However, information about individual virtual channels within the virtual path is not required; all the virtual channels of that virtual path follow the same route as the virtual path. In certain switching nodes (VC switch and VC cross-connect), VPI translation is provided in conjunction with the VCI translation mentioned above.

At the virtual path level, VPCs are provided for the purpose of *user-to-user, user-to-network* and *network-to-network* information transfer in a similar way to that described for VCCs. A possible example for each of the three cases is shown in Figure 2.9, but it must be emphasised that other uses for the different types of configuration exist and that the purpose of the figure is to *illustrate* the principle.

Figure 2.9a shows a user-to-user example. This is equivalent to a leased line and the customer can use the virtual path for any mix of applications, provided that the parameters (for example, available bit-rate, quality of service) of the virtual path are adequate. The network will only consider the parameters of the VPC when it is established and will only apply traffic control to the virtual path, not to the virtual channels within it. Using virtual paths in this way provides a much more flexible approach to building a private network than does the use of physical leased lines. Since the virtual paths are set up by network management they can be made available on a *reserved* basis (perhaps at certain times of the day) rather than being permanent.

User-to-network applications (Figure 2.9b) allow a customer to have separate VPIs for groups of connections that are to be routed to different networks or service providers. Another use of such a user-to-network arrangement would be to provide a different quality of service with each virtual path.

In the public network, a possible use of a network-to-network virtual path is to provide a dedicated traffic route between two exchanges that have a high traffic load between them (Figure 2.9c). This enables the connections to be switched through intermediate exchanges at a virtual path level without the intermediate exchange having to be aware of VCCs. Because the VPCs are established by network management, they can be changed to reflect varying traffic demand.

The general properties of VCCs are also applicable to VPCs; usage parameter control is provided on a virtual path basis as well as a VC basis. However, not all the VCIs theoretically available within a VPC may be available for the user of the VPC, some being reserved for other purposes.

Establishment and release of VPCs may be performed by signalling or by network management procedures, depending on whether the VPC is to be provisioned or made available on demand. On-demand set-up and release may, in principle, be done by the network or by the customer.

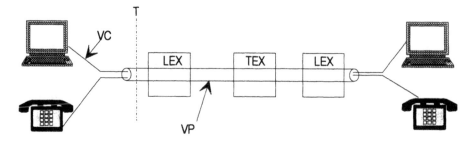

a) user-to-user VP: the network is only concerned with routeing and controlling the VP, not the individual VCs within it

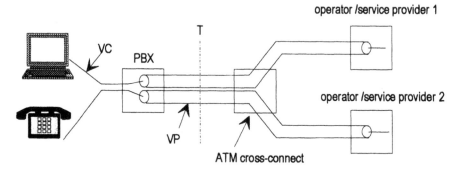

b) user-to-network VP: a possible application is enabling the customer to have access to two different network operators or service providers

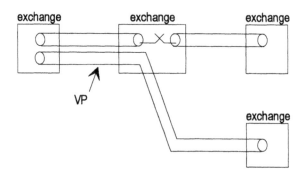

c) a use of network-to-network VPs would be to provide direct traffic routes without intermediate VC switching where there is sufficient traffic demand

Figure 2.9 Possible applications of VPs with different locations of endpoint

2.5. ATM adaptation layer

2.5.1. *AAL principles*

The ATM adaptation layer (AAL) performs the necessary mapping between the ATM layer and the next higher layer. This is done in the terminal equipment or terminal adapter, i.e. at the edge of the ATM network.

The ATM network, the part of the network which processes the functions of the ATM layer, is independent of the telecommunications services it carries. This means that the user payload is carried transparently by the ATM network and the ATM network does not process the user payload and does not know the structure of the data unit. This is known as *semantic independence*. There is also *time independence* as there is no timing relationship between the clock of the application and the clock of the network, the network having to cope with any application bit-rate.

The consequence of this independence is that all the functions specific to the services are provided at the boundary of the ATM network and are performed by the AAL. The functions within the AAL have the task of providing the data flow sent by the user to the upper layers at the receiving end, taking into account any effects introduced by the ATM layer. Within the ATM layer, the data flow can be corrupted by errors in transmission or it can suffer cell delay variation as a result of variable delay in buffers or through congestion in the network. Loss of cells or misdelivery of cells are consequences of these effects and these consequences will have an impact on the application, whether it is data transfer, video or voice communication. The AAL protocols must cope with these effects. It could be thought that for each telecommunications service there should be a separate AAL developed. However, taking into account the common factors within possible telecommunications services it is possible to devise a small set of AAL protocols that should be sufficient for what is currently envisaged.

The functions performed in the AAL depend upon the higher-layer requirements and, as the AAL supports multiple protocols to fit the needs of the different AAL service users, it is therefore service dependent. It is obviously sensible to minimise the number of different AAL protocols required and, to do this, a telecommunication service classification is defined based on the following parameters:

- Timing relationship between source and destination
- Bit-rate
- Connection mode.

Other parameters, such as assurance of the communication, are treated as quality-of-service parameters and do not lead to different classes.

Using these parameters, four *classes* of service have been defined and these are shown in Table 2.3. Examples of telecommunication services in each of the classes are:

- Class A: circuit emulation, constant bit-rate video
- Class B: variable bit-rate video and audio
- Class C: connection-oriented data transfer
- Class D: connection-less data transfer

Table 2.3   Classes of service

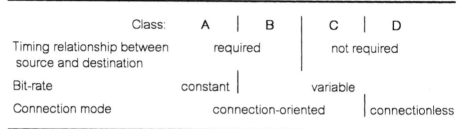

Class:	A	B	C	D
Timing relationship between source and destination	required		not required	
Bit-rate	constant		variable	
Connection mode	connection-oriented			connectionless

These *classes of service* are a fairly general concept, but the classes can be mapped onto the different specific *AAL types* as shown in Table 2.4. In addition, there is an empty AAL for those users for whom the performance of the ATM layer may be sufficient for their requirements.

Table 2.4   Mapping between classes of service and AAL types

class	
A:	AAL 1
B:	AAL 2 (not yet defined)
C & D	AAL 3/4 (the original AAL 3 and AAL 4 combined)
C & D	AAL 5 is alternative to AAL 3/4 for class D services; it is simpler and therefore has less overhead

The AAL is organised in two sublayers (Figure 2.10): the *convergence sublayer* (CS) and the *segmentation and reassembly sublayer* (SAR).

The SAR layer is concerned with the segmentation of higher-layer information into a suitable size for the information field of an ATM cell and for reassembly of the contents of ATM cell information fields into higher-layer information.

This is not enough to reconstitute the information sent and other functions have to be performed. These include processing of cell-delay-variation, end-to-end synchronisation, handling of loss and misinserted cells. Such functions are performed by the convergence sublayer. This sublayer is service specific and different convergence sublayers may be used on top of the same SAR.

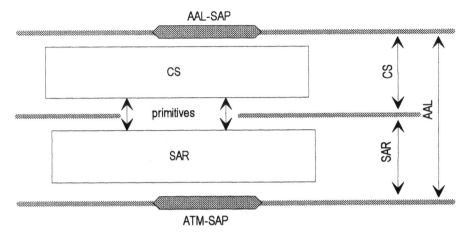

Figure 2.10 General structure of AAL

The functions of the AAL may be empty if the performance of the ATM layer is sufficient for the requirements of that particular telecommunications service. In this case, users are able to make use of all 48 octets in the information field. The addition of headers and trailers in each sublayer, together with the segmentation of user information to fit the ATM cell, is shown in Figure 2.11.

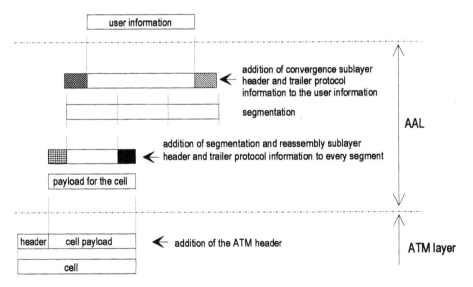

Figure 2.11 Mapping of user information into an ATM cell through the AAL

For specific needs in two AAL types, the convergence sublayer has been subdivided into two parts (Figure 2.12), the *common part CS* (CPCS) and the *service specific CS* (SSCS). Different protocols, to support specific AAL user services, or group of services, may be defined. The SSCS may be null, in the sense that it only provides for the mapping of the equivalent primitives of the AAL to CPCS and vice-versa.

Definitions of the functions and protocols within the AAL follow the usual naming conventional associated with layered systems [65] with *protocol data units* (PDUs) and *service data units* (SDUs). The SDU passes across *service access points* (SAPs) whereas the PDU is the data unit between peer layers. For ease of reference this is shown diagrammatically as it applies to the AAL in Figure 2.13. For AAL type 3/4 and type 5, which both have the SSCS and CPCS, there is a PDU defined for each of this parts, with the appropriate header and trailer. Because no service access point is defined between the CS and SAR sublayers, there is no 'SAR SDU'.

The definition of all the functions and protocols within the AAL is a complex process and, at the time of writing this book, they have not all been completely defined within CCITT. The following sections, therefore, highlight the main features of the AAL types.

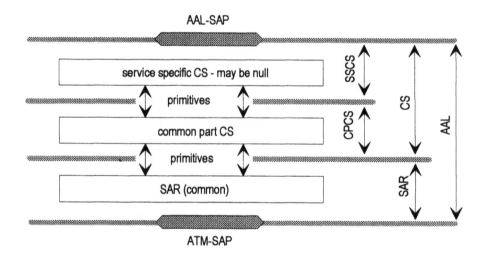

Figure 2.12 Structure of AAL with SSCS and CPCS (used in AALs 3/4 and 5)

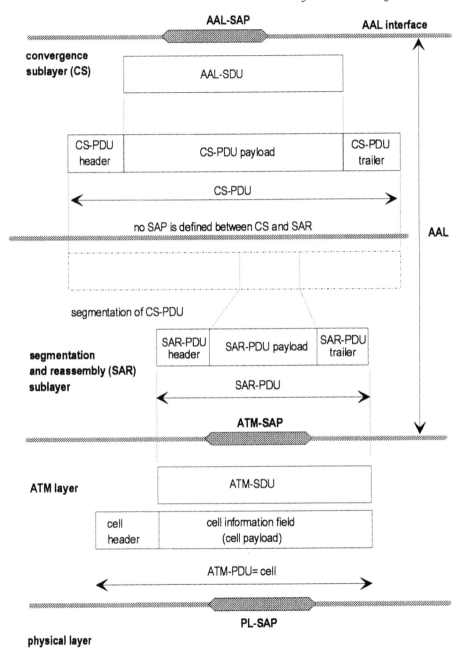

Figure 2.13 General data unit naming convention

2.5.2.   AAL type 1

AAL 1 provides the following layer services to the AAL user:

- Transfer of service data units with a constant source bit-rate and their delivery with the same bit-rate
- Transfer of timing information between source and destination
- Transfer of structure information between source and destination
- Indication of lost or errored information which is not recovered by AAL 1, if needed.

The functions listed below may be performed in the AAL in order to enhance the layer service provided by the ATM layer. The SAR-PDU format is given in Figure 2.14.

- Segmentation and reassembly of user information
- Handling of cell delay variation
- Handling of cell payload assembly delay
- Handling of lost and misinserted cells
- Source clock frequency recovery at the receiver
- Recovery of the source data structure at the receiver
- Monitoring of AAL-PCI (protocol control information) for bit errors
- Handling of AAL-PCI bit errors
- Monitoring of the user information field for bit errors and possible corrective action.

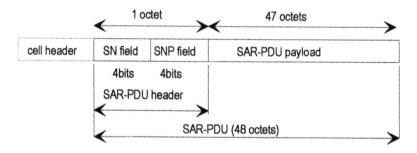

SN sequence number for numbering of the SAR-PDUs
SNP sequence number protection to protect the SN field

Figure 2.14 SAR PDU format of AAL 1

At the SAR level, the SNP field, which provides 1-bit error correction and 2-bit error detection, is processed. If the result is right (no error detected or error detected and corrected), the SN field is sent to the CS level, which processes it depending on the application. Four CS have been identified for the following applications:

- Circuit transport to support both asynchronous and synchronous circuits. Examples of asynchronous circuit transport are 1 544, 2 048, 6 312, 8 448, 32 064, 34 368 and 44 736 kbit/s. Examples of synchronous circuit transport are signals at 64, 384, 1 536 and 1 920 kbit/s
- Video signal transport for interactive and distributive services
- Voice-band signal transport
- High-quality audio signal transport.

2.5.3. AAL type 2

At the end of the CCITT study period 1989-1992, the AAL 2 had not been defined. However, there has been an indication of possible requirements and the services provided by AAL 2 to the higher layer may include:

- Transfer of service data units with a variable source bit-rate
- Transfer of timing information between source and destination
- Indication of lost or errored information which is not recovered by AAL 2.

2.5.4. AAL types 3/4

AAL 3 was designed for Class C services (connection-oriented data) and AAL 4 for Class D services (connectionless-oriented data). During the standardisation process, the two AALs were merged and are now the same. The current name given to this AAL is AAL 3/4.

Two modes of service have been defined: *message* and *streaming*. These are explained below.

In *message mode* (Figure 2.15), the AAL service data unit is passed across the AAL interface in exactly one AAL interface data unit (AAL-IDU). This service provides the transport of fixed length or variable length AAL-SDUs.

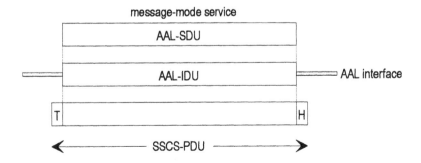

Figure 2.15 Message-mode service

In *streaming mode* (Figure 2.16) the AAL service data unit is passed across the AAL interface in one or more AAL interface data units. The transfer of these AAL-IDUs across the AAL interface may occur separated in time and this service provides the transport of variable length AAL-SDUs. The streaming-mode service includes an abort service by which the discarding of an AAL-SDU partially transferred across the AAL interface can be requested.

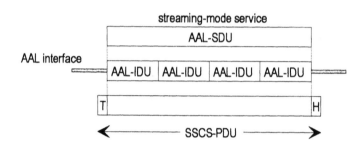

Figure 2.16 Streaming-mode service

2.5.4.1. *The segmentation and reassembly sublayer (SAR)*

The SAR sublayer is depicted in Figure 2.17. The SAR sublayer accepts variable-length CS-PDUs from the convergence sublayer and generates SAR-PDUs with a payload of 44 octets, each containing a segment of the CS-PDU.

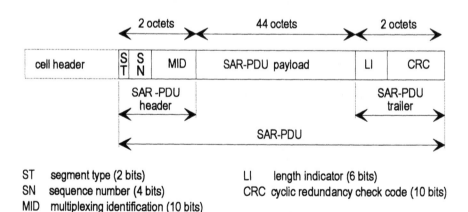

ST   segment type (2 bits)   LI   length indicator (6 bits)
SN   sequence number (4 bits)   CRC cyclic redundancy check code (10 bits)
MID   multiplexing identification (10 bits)

Figure 2.17 SAR format of AAL 3/4

The *segment type* indication identifies a SAR-PDU as containing a *beginning of message*, a *continuation of message*, an *end of message* or a *single segment message*. The *sequence number* allows the sequence of SAR-PDUs to be numbered modulo 16.

These two fields enable the segments of the CS-PDU to be re-assembled in the correct sequence and minimise the effect of errors on the reassembly process.

The *multiplexing identification* is used to identify a CPCS connection on a single ATM-layer connection. This allows for more than one CPCS connection for a single ATM-layer connection. The SAR sublayer, therefore, provides the means for the transfer of multiple, variable-length CS-PDUs concurrently over a single ATM layer connection between AAL entities.

The *length indication* contains the number of octets of CS-PDU information that are included in the SAR-PDU payload field. This is necessary because the amount of data from the CS-PDU may not completely fill the 44 octets available.

The *CRC* field is a 10-bit sequence used to detect bit errors across the whole SAR-PDU. This includes the CS-PDU segment and hence the user data.

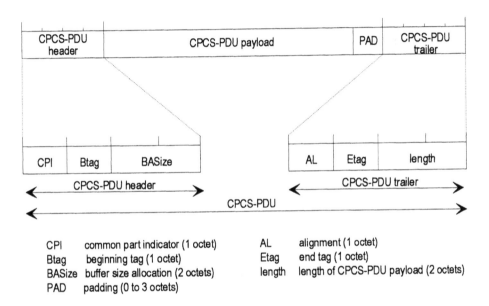

Figure 2.18 CPCS-PDU format for AAL 3/4

2.5.4.2. The convergence sublayer (CS)

The CS has been subdivided into the *common-part CS* (CPCS) and the *service-specific CS* (SSCS). At the time of writing, the SSCS has not yet been defined so it will not be considered further.

The functions of the CPCS are:

- Preservation of SSCS-PDUs
- Error detection and handling
- Buffer allocation size
- Abort

The CPCS requires a 4-octet header and a 4-octet trailer. The padding field provides a 32 bit alignment of the CPCS-PDU payload. The sizes and position of fields for the CPCS-PDU structure are given in Figure 2.18.

2.5.5.   AAL type 5

This new AAL was introduced in the study process of CCITT at the end of 1991. Its description is almost complete and will be published in the 1994 CCITT recommendations. Designed for the same class of service as AAL 3/4, it has the advantage of being simpler and requiring less overhead. Unlike AAL 3/4, it allows all 48 octets of the cell information field to be used for the transport of CS-PDU segments, the only SAR protocol information being provided by a bit in the ATM cell header, as explained below. This means that there is neither multiplexing nor error control at the SAR sublayer. However, there is a CRC field in the CS sublayer.

There are also similarities with AAL 3/4. The two modes of service defined, message and streaming, are the same as those for AAL 3/4. Also similar to AAL 3/4, the convergence sublayer of AAL 5 has been subdivided into a CPCS part and a SSCS part. Figure 2.19 shows the SAR format and Figure 2.21 the CS.

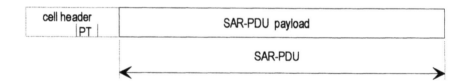

PT: payload type: the PT belongs to the ATM header and it conveys the value of the
ATM-layer-user-to-ATM-layer-user indication.

Figure 2.19 SAR format of AAL 5

The protocol control information of the SAR sublayer uses the ATM-layer-user-to-ATM-layer-user parameter (AUU) contained in the ATM header to indicate that a SAR-PDU contains the end of a CS-PDU. When this bit is set to 1, it indicates the end of the CS-PDU; when set to 0 it indicates the continuation or the beginning of a CS-PDU. This is necessary to enable the SAR to cope with reassembly of the CS-PDU in the presence of errors. If no indication of the end of the CS-PDU was provided, the loss of a cell, and hence the loss of a segment of the CS-PDU, would mean that all subsequent reassembly operations would be incorrect. By indicating the end of the CS-PDU, the loss of a single cell would limit the error to one CS-PDU, unless the lost cell contained the end indication in which case the error would be restricted to 2 CS-PDUs. This is illustrated in Figure 2.20.

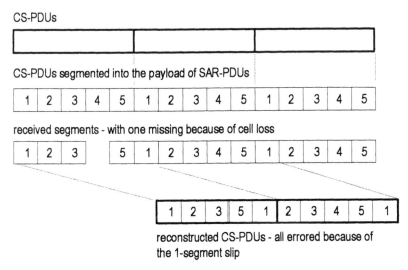

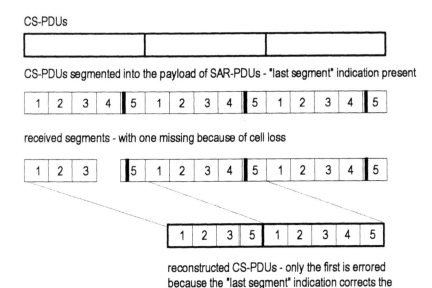

a) without "last-segment" indication

b) with "last-segment" indication

Figure 2.20 Illustration of the need for segment marking to correctly delineate CS-PDUs

The format of the CPCS is shown in Figure 2.21. The main CPCS functions are preserving the SSCP-PDU, providing CPCS user-to-user indication, detecting and handling errors, providing an 'abort' function and also padding where necessary.

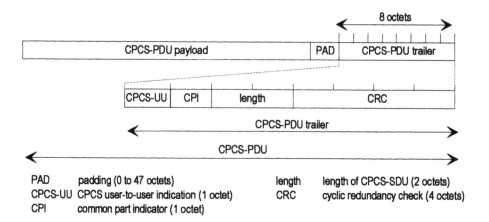

Figure 2.21 CPCS-PDU format for AAL 5

2.5.6.   Service timing/synchronisation aspects

Service requirements for timing functions vary widely and may be supported in a number of ways, based both on end-to-end service information and on facilities available from the network. Some existing services with 8 kHz integrity (such as voice telephony) will require network-provided facilities. New services may need to use end-to-end techniques to meet performance requirements. In addition, a combination of both methods may be used.

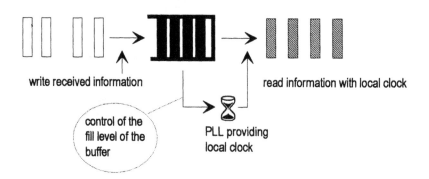

Figure 2.22 Adaptive clock at the receiving side

Two examples of end-to-end service timing methods, which may be suitable for services such as asynchronous CBR services, are described below. They may be included in the service specification as appropriate to meet the performance requirements of the service:

The first example uses an *adaptive clock* (Figure 2.22). The receiver writes the received information field into a buffer and then reads it with a local clock. The filling level of the buffer is used to control the frequency of the local clock. The control is performed by continuously measuring the fill level around its median position, and by using this measure to drive the phased-locked loop providing the local clock.

The second example uses the *synchronous residual time stamp* method (Figure 2.23). This method measures and conveys information about the frequency difference between a common reference clock derived from the network and the service clock. The same network clock is assumed to be available at both the transmitter and the receiver. The time stamp technique is well suited to eliminate very low frequency jitter problems and a typical use is in the transport of G.702 signals [18] to meet jitter performance specified in CCITT Recommendations G.823 and G.824.

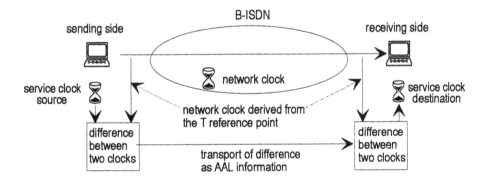

Figure 2.23 Use of synchronous residual time stamp method

2.6. Physical layer

2.6.1. *Principles*

The physical layer consists of two sublayers: the *physical medium* sublayer (PM) and the *transmission convergence* sublayer (TC).

The physical medium sublayer includes only physical-medium-dependent functions and provides bit transmission capability, including bit-transfer and bit-alignment. It includes line coding and electrical-optical transformation.

The transmission convergence sublayer performs all those functions necessary to transform a flow of cells into a flow of data units (e.g. bits) which can be transmitted and received over a physical medium.

Going from the physical layer to the ATM layer, the flow of data (strictly, in OSI terms, the flow of *service data units*) crossing this boundary is a flow of valid cells (this was shown earlier in Figure 2.5). Valid cells are those whose headers have no errors, error checking on the header having been performed in the transmission convergence sublayer.

Going in the opposite direction, from the ATM layer to the physical layer, the ATM cell flow is merged with the appropriate information for cell delineation and it also carries operation and maintenance (OAM) information relating to this cell flow.

2.6.2.   Transfer capacity

CCITT Recommendation I.432 defines two bit-rates for the physical layer at the T_B reference point: 155 520 kbit/s in both directions and 622 080 kbit/s in at least one direction. The interfaces may be electrical or optical and it may use a cell-based structure or SDH framing.

The bit-rates above are the gross bit-rates of the physical layer and the overhead due to physical-layer frame-structure octets or physical-layer cells must be subtracted. The values available for ATM cells are shown in Table 2.5 and these correspond to the payload of SDH. The maximum user bit-rate is 48/53 of that available for cells and this too is shown in Table 2.5.

However, the bit-rate available for ATM cells must also be used for signalling cells and OAM information cells for the ATM and higher layers. The bit-rate available for user information is consequently less than the values in the table.

Table 2.5   Physical layer parameter values

	lower bit-rate (kbit/s)	higher bit-rate (kbit/s)
gross physical-layer bit-rate	155 520	622 080
max bit-rate available for ATM cells	149 760	599 040
max bit-rate available for cell payload	135 631	542 526

2.6.3.   Physical layer for cell-based interface

The interface structure consists of a continuous stream of cells. The maximum spacing between successive physical-layer cells is 26 ATM-layer cells, i.e. after 26 contiguous ATM-layer cells have been transmitted, a physical-layer cell is inserted in order to adapt the transfer capability to the interface rate. Physical-layer cells are also inserted when no ATM-layer cells are available. The physical-layer cells which are inserted can be either 'idle cells' or physical-layer OAM cells, depending on the OAM requirements.

Physical-layer OAM cells are used for the conveyance of the physical-layer OAM information. How often these cells are inserted should be determined by OAM requirements. However, there must be not more than one physical-layer OAM cell every 27 cells and not less than one physical layer OAM cell every 513 cells on an operating link.

The physical-layer OAM cells have a unique header so that they can be properly identified by the physical layer.

2.6.4. *Physical layer for SDH-based interface*

For the 155.52 Mbit/s interface, the ATM cell stream is first mapped into the container C-4 and then mapped in the VC-4 virtual container along with the VC-4 path overhead (Figure 2.24). The ATM cell boundaries are aligned with the STM-1 octet boundaries. Since the C-4 capacity (2 340 octets) is not an integer multiple of the cell length (53 octets), a cell may cross a C-4 boundary.

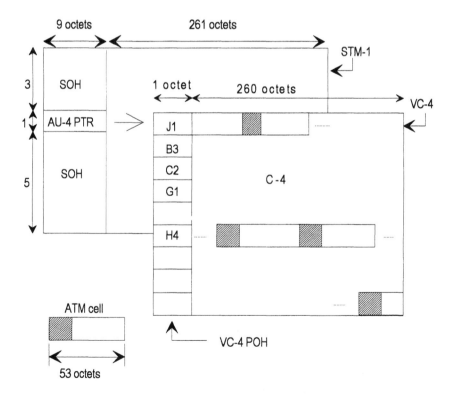

Figure 2.24. Structure of 155 520 kbit/s frame for SDH-based UNI

At 622 080 kbit/s, the ATM cell stream is first mapped into the container C-4-4c and then packed in the VC-4-4c virtual container along with the VC-4-4c path overhead. The ATM cell boundaries are aligned with the STM-4 octet boundaries. Since the C-4-4c capacity (9 360 octets) is not an integer multiple of the cell length (53 octets), a cell may cross a C-4-c boundary.

2.6.5. Header error control

The header error control (HEC) covers the entire cell header. The code used for this function is capable of either single-bit error correction or multiple-bit error-detection.

The transmitter calculates the HEC value (an 8-bit sequence) across the entire ATM cell header and inserts the result in the HEC field. The value is the remainder of the division (modulo 2) by the generator polynomial $x^8 + x^2 + x + 1$ of the product x^8 multiplied by the content of the header *excluding* the HEC field.

To significantly improve the cell delineation performance in the case of bit slips, the CCITT recommends that the check bits calculated by the use of the check polynomial are added (modulo 2) to an 8-bit pattern (01010101, the left bit being the most significant) before being inserted in the HEC field. The receiver must subtract (equal to add modulo 2) the same pattern from the 8 HEC bits before calculating the syndrome of the header. This operation does not affect the error detection/correction capabilities of the HEC.

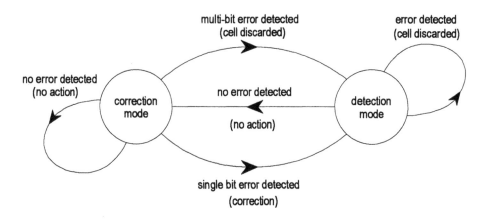

Figure 2.25 HEC: receiver modes of operation

The receiver has two modes of operation (Figure 2.25). The default mode provides for single-bit error correction. Each cell header is examined and, if an error is detected, one of two actions takes place depending on the state of the receiver. In 'correction mode', only single-bit errors can be corrected and the receiver switches to 'detection mode'. In 'detection mode', all cells with detected header errors are discarded. When a header is examined and found not to be in error, the receiver switches to 'correction mode'.

The error-protection function of the HEC provides both recovery from single-bit header-errors, and a low probability of the delivery of cells with errored headers under bursty error conditions. The error characteristics of fibre-based transmission systems appear to be a mix of single-bit errors and relatively large burst errors.

2.6.6. Cell delineation and scrambling

Cell delineation is the process which allows identification of the cell boundaries and is performed using a procedure based on the header error control (HEC) field. Scrambling is used to improve the security and robustness of the HEC cell delineation mechanism. In addition it helps to randomise the data in the information field for possible improvement of the transmission performance. Any scrambler specification must not alter the ATM header structure, header error control and cell-delineation algorithm.

The recommended cell delineation method is performed by using the correlation between the header bits to be protected (32 bits) and the relevant control bits (8 bits) introduced in the header by the HEC (header error control) using a shortened cyclic code with generating polynomial $x^8 + x^2 + x + 1$.

Figure 2.26 shows the state diagram of the HEC cell delineation method.

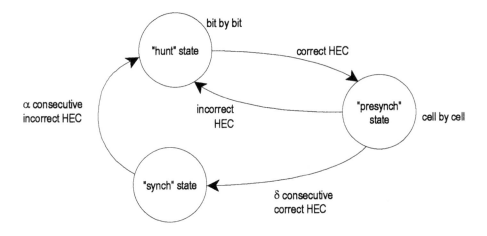

Figure 2.26 Cell delineation state diagram

- In the *hunt* state, the delineation process is performed by checking bit by bit for the correct HEC (i.e. syndrome equals zero) for the assumed header field. For the cell-based physical layer, prior to scrambler synchronisation, only the last six bits of the HEC are used for cell delineation checking. Once such an agreement is found, it is assumed that one header has been found, and the method enters the *presynch* state. When octet boundaries are available within the receiving physical layer prior to cell delineation, the cell delineation process may be performed octet by octet.

- In the *presynch* state, the delineation process is performed by checking cell by cell for the correct HEC. The process repeats until the correct HEC has been confirmed δ times consecutively. If an incorrect HEC is found, the process returns to the *hunt* state.
- In the *synch* state the cell delineation will be assumed to be lost if an incorrect HEC is obtained α times consecutively.

The parameters α and δ have to be chosen to make the cell delineation process as robust and secure as possible and able to satisfy the required performance. Robustness against false misalignments due to bit errors depends on α; robustness against false delineation in the re-synchronisation process depends on the value of δ. The values suggested by the CCITT in Recommendation I.432 are shown in Table 2.6.

Table 2.6 Values of α and δ from CCITT

	α	δ
SDH-based physical layer	7	6
cell-based physical layer	7	9

2.6.7.   Scrambler operation

For the cell-based UNI, the *distributed sample scrambler* is recommended. This is an example of a class of scrambler in which randomisation of the transmitted data stream is achieved by modulo addition of a pseudo-random sequence. Descrambling at the receiver is achieved by modulo addition of an identical locally generated pseudo-random sequence having phase synchronisation with the first in respect of the transmitted cells. The scrambler does not affect the performance of the 8-bit HEC mechanism during steady-state operation.

For the SDH-based physical layer, CCITT has proposed the following polynomial: self-synchronising scrambler $x^{43} + 1$. This has been selected to minimise the error multiplication (two) introduced by the self synchronised scrambler and is described in detail in CCITT Recommendation I.432.

Chapter 3

Applications and Services

3.1. Applications

Although this book is mainly concerned with the network aspects of ATM, some regard must be paid to the applications that can make use of a broadband communications infrastructure. It cannot be emphasised too strongly that the customer will pay for services only if these services support the required application and are seen to be priced at a level that makes their use worthwhile.

When considering applications it is important to separate applications and services:

- A *service* is what the customer pays for
- An *application* is how a customer makes use of a service
- A *teleservice* is one of standardised set of services

For instance, the connecting together of two LANs is an *application* that could make use of a special *LAN interconnection service* or a *transparent VP service*.

A detailed treatment of possible applications and services is not appropriate for a book of this nature, but the following examples represent the type of application that the network must cater for:

For business customers

- High-speed LAN-to-LAN interconnection offers the ability to access large data files on remote servers. Although applications such as computer aided design come readily to mind, there are many others in the world of commerce and business. One proposed application is in the insurance business, where clerks have access from their local terminals to remote databases. Records, often large files containing scanned images of insurance claims, need to be accessed quickly, especially when the clerk is responding to a telephone query. If these records are held on centralised databases, access rates need to be high so that the insurance company can give a satisfactory service to its customers.

- Of interest to customers with large computing centres is disaster recovery: alternative computing centres are kept on 'hot stand-by' to take over from main sites. Thus, in case of fire or any other disaster, computer operations are not lost, traffic being switched to the secondary site with little effect on the users.

- In the medical field, B-ISDN could be used for the transmission of large data files containing medical images such as those created by tomographic imaging and digitised X-ray images. This would allow a surgeon to confer with a colleague in a remote hospital to gain a second opinion on a medical condition. Combining the file transmission with a video-conference would allow opinions to be gained quickly and hence improve medical care.

- 'Multimedia' conferencing is expected to be a large growth area, as it would allow such techniques as the joint editing of documents within a work team as well as video-conferencing. Each member of the group would have one window open on a workstation screen as a video-phone, another window for joint editing, another as a scratchpad area and so on. The reason that this is likely to be such an attractive application for business is that it will reduce the need for travel. It is not just the cost of fares and hotels that makes business travel so expensive, but the cost of the unproductive time.

- Other video-conferencing-based applications are in the field of training. Industrial companies require a large amount of training to keep their employees up to date with the latest innovations and this is often done by sending employees to training centres or by showing video tapes at the workplace. Video-conferencing could be used to provide interactive teaching from a remote site to the employees at their own workplace or a closely-located training centre.

For residential customers

- There have been some interesting studies across Europe in the field of social services, looking at providing support at home for the elderly, together with broadband interactive video links to keep the old people in touch with their family as an alternative to residential care. The rationale behind this work is that it is often loneliness, usually resulting from the death of a spouse, that leads to elderly people no longer taking care of themselves and so ending up in residential care.

- Tele-working (sometimes called 'tele-commuting') is a 'green' application of B-ISDN. It enables someone to work at home instead of travelling to the office everyday, so reducing the impact of that travelling on the environment. There are many office jobs that can be done equally as well at home or in a local centre (often called a 'tele-cottage'), provided that the worker has access to a good computing and communications infrastructure. Home working can be attractive to the employer, who does not have to provide expensive office accommodation, and to those employees who have domestic commitments or who live in remote regions. The drawback is that the employee can feel isolated and miss the social interaction of an office environment and it is here that the video capability of B-ISDN helps to retain contact.

- Video-retrieval is an oft-quoted example of a B-ISDN application for residential customers: the customer can retrieve a video from a video library on demand. The challenge for such a service is to provide it a price that is competitive with the very low cost of renting a video cassette.

- Tele-learning is a similar idea to the use of video-conferencing for training in the business community. It enhances the existing uses of video for home study by providing interactive video links to enable the students to question the teacher.

- An application that combines both video-retrieval with transaction processing is tele-shopping. This is a video equivalent of catalogue shopping, whereby the customer can have on-demand access to a video of the chosen product range. The video provides more information than can be given in the still photograph in a catalogue, and when the desired product has been chosen it can be ordered and paid for interactively.

The problem for the network operators when considering the introduction of new services is that it is very difficult to produce a realistic forecast of demand. There is little or no information available on which to base the assumptions of demand and there have been memorable occasions in the past when new services have failed to generate sufficient demand and have been abandoned.

3.2. Classification of services

In order to be able to define suitable attributes for teleservices and bearer services, as well as to plan the network in general, it is necessary to have an understanding of the services to be carried over the network. With a multitude of possible applications and corresponding services, it is obviously helpful to be able to have some broad classification. CCITT has done this in Recommendation I.211, producing two broad categories (interactive and distribution services), each of which is divided into service *classes* (Table 3.2). Sometimes the term 'distributive' is seen instead of 'distribution'.

Conversational services are generally bi-directional and real-time: there is no store-and-forward mechanism. These are perhaps the easiest types of service to visualise, as they correspond more closely to the common types of service in existing telecommunications networks. Examples of broadband conversational services are video-telephony and video-conferencing. Many uses of data transfer (such as terminal access) are also real-time and interactive and so fall into this class.

Table 3.1   General classification of services

Interactive services	conversational services messaging services retrieval services
Distribution services	distribution services *without* user presentation control distribution services *with* user presentation control

Messaging services imply the use of store-and-forward mechanisms which may also allow some form of processing functions. Computer users will be very familiar with electronic mail, which is an example of a current messaging service; another is voice mail. In the broadband environment it will also be possible to mail moving video images.

With retrieval services, the information is stored in a central library and transmitted to a particular user on request. A possible broadband application is a library from which the user would be able to retrieve video information (e.g. a film) or audio information (perhaps a recording of a concert) on demand.

Distribution services provide a continuous flow of information from a central source to an unlimited number of users. Without user presentation control, an individual user can start to access this information whenever he or she wishes but cannot control the start of the information transmission. Broadcast television is a good example of such a service. When the service offers user presentation control, the information from the central source is transmitted as a cyclical repetition of information entities (for examples frames or pictures) and the user can control his or her access to the information flow so that entities selected will always appear from the beginning of that sequence. Teletext is an example of a current service of this type; with broadband it could be extended to sequences of moving images.

3.3. Teleservices

In the CCITT Recommendation I.112 the term *teleservice* is defined as follows:

> 'A type of telecommunications service that provides the complete capability, including terminal equipment functions, for communication between users according to protocols established by agreement between Administrations'

A teleservice description characterises the terminal and network combination that fulfils the service requirements seen by the user. Normally, there are several terminal/network combinations for the same user service. The user sees a telecommunications service at the *application layer* (layer 7), but the definition of a teleservice covers all layers of the OSI model. The bearer service is the service provided by the *network layer* (layer 3) to the higher layers and is concerned only with layers 1-3.

The teleservice is, therefore, a particular application running over a particular set of low layers with the combination accepted by the network operator. Using the same set of higher layers on a different set of low layers would produce a different teleservice as shown in Figure 3.1.

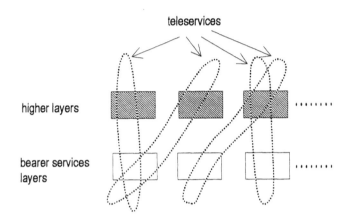

Figure 3.1 The association between teleservices and bearer services

Teleservices are characterised by a set of *low-layer attributes, high-layer attributes* and some *general attributes.* The low-layer attributes are those used to characterise the bearer capability and are described in detail in the section on bearer services. CCITT Recommendation I.210 states that the high-layer attributes refer to the functions and protocols of layers 4-7 of the OSI model.

High-layer attributes

- Type of user information (e.g. speech, facsimile)
- Layer 4 (transport layer) protocol functions
- Layer 5 (session layer) protocol functions
- Layer 6 (presentation layer) protocol functions
- Layer 7 (application layer) protocol functions

General attributes (these have not yet been fully defined within CCITT)

- Supplementary low and high-layer attributes (supplementary services)
- Quality of service (user-oriented)
- Interworking possibilities
- Operational and commercial

3.4. Bearer services

3.4.1. Definition

The term bearer service is defined in CCITT Recommendation I.112 as:

> 'A type of telecommunication service that provides the capability for the transmission of signals between user-network interfaces.'

The description of a bearer service only covers aspects related to the transfer of information between UNIs, because, in principle, it can support any number of teleservices, depending on the higher layers.

Bearer services are characterised by the CCITT using a set of independent low-layer salient features, as well as a set of more general features, called attributes, a particular bearer service being identified by a unique set of values for these attributes. The user of a bearer service may choose any set of higher-layer (at least layers 4-7) protocols for the communication and the B-ISDN does not ascertain compatibility between customers at these layers.

Bearer service attributes are currently classified into three categories as follows:

Information transfer attributes (low-layer)

- Information transfer mode
- Information transfer rate
- Information transfer capability
- Structure
- Establishment of communication
- Symmetry
- Communication configuration

Access attributes (low-layer)

- Access channel and rate
- Access protocols

General attributes

- Supplementary services provided
- Quality of service (network-oriented)
- Interworking possibilities
- Operational and commercial aspects

These attributes and their values were defined in a 64 kbit/s environment (N-ISDN) and, in order to take account of the broadband network and the ATM transfer mode, new attributes or sub-attributes and new values of the existing attributes have to be defined. These are considered in the next sections.

3.4.2.   Information transfer mode

This attribute describes the operational mode for transferring (by transport and switching) user information through a B-ISDN network. The value of this attribute is *ATM*.

Four sub-attributes characterising the service requirements on an ATM network have also been added:

- Connection mode: describing a *connection oriented* or a *connectionless transfer*
- Traffic type (within the virtual channels or virtual paths): describing bit streams with *constant* or *variable bit-rates* supported by the service
- Timing end-to-end of virtual channels and virtual paths: describing the timing relationship between source and destination of signals. End-to-end timing is necessary for real-time information, e.g. video signals
- VCI transparency for virtual-path services: this sub-attribute is relevant for virtual-path bearer service. It describes the unrestricted use and multiplexing of virtual channels within the VP by the users.

3.4.3.   Information transfer rate: bit-rate parameters

It is a feature specific to ATM that there is no particular information transfer rate: in principle a service could use any bit-rate up to the maximum capacity of the link. A significant amount of work has been done to find parameters that are simple enough to be monitored and controlled (sometimes called 'policed') and detailed enough to allow for an efficient connection-admission-control function exploiting the statistical multiplexing gain as far as possible.

The bit-rate of variable-bit-rate services available to the user for the transfer of user's information via one or several VCs or VPs is mainly described by two sub-attributes:

- *Peak bit-rate*: describing the maximum bit-rate offered to the user for a given time period for the transfer of a bursty signal
- *Mean bit-rate*: describing the average bit-rate available to the user for a given time period.

The bit-rate offered by a constant-bit-rate service will be characterised by the sub-attribute *peak bit-rate*.

The source traffic may be further described by additional attributes, such as 'burstiness' and 'peak duration'. This area is still under study and it is not completely clear yet what, if any, additional sub-attributes will be used.

3.4.4. Information transfer capability of VC and VP

This attribute describes the capability associated with the transfer of different types of information. The value used to characterise an ATM bearer service should be *unrestricted digital information*. When interworking with other networks, the value of this attribute corresponds to the value of the attribute in the other network. Typical values when interworking could be *speech, audio,* etc.

3.4.5. Structure of VC and VP

This attribute refers to the capability of B-ISDN to deliver information to the destination in a structure corresponding to that presented at the origin. For a bearer service in an ATM environment, the value should be *AAL service data unit integrity* (48 octets in the case of the empty AAL).

3.4.6. Establishment of communication

This attribute is associated with a telecommunication service and describes the mode of establishment and release of a communication. A bearer service in an ATM environment must provide all values required by the various telecommunication services. Possible values are *on-demand, reserved* and *permanent.*

On demand means that the connection can be established or released as soon as possible after the request is made.

A *reserved* communication is set up for a specified period of time; the times at which establishment and release of the communication being predetermined. As an option, a release request can be accepted during the communication.

A *permanent* communication is set up in response to a subscription request for the service. The duration may be unspecified and the communication is released at a time corresponding to the end of the subscription.

3.4.7. Symmetry

This attribute describes the symmetry relationship of information flows between two access points involved in a communication. Possible values are *unidirectional, bi-directional symmetric* and *bi-directional asymmetric.*

The value *unidirectional* applies when the information flow is provided only in one direction.

If the information flow characteristics are the same between the two access points in the forward and backwards direction, the value is *bi-directional symmetric.*

The value *bi-directional asymmetric* should be used if the information-flow characteristics are different in the two directions.

3.4.8. Communication configuration

This attribute describes the spatial arrangement for transferring information between two or more network terminations or reference points. A bearer service in an ATM environment must provide all the configurations required by the teleservices. Possible values for the attribute are *point-to-point*, *point-to-multipoint* and *broadcast*.

Point-to-point communication is from a single point to another single point. *Point-to-multipoint* communication is from a single point to a limited number of specified destinations. *Broadcast* communication is from a single point to an unlimited number of unspecified destinations.

Choosing a particular communications-configuration attribute will have an impact on the choice of other attributes. For example, the choice 'broadcast' will limit the choice of symmetry attribute to 'unidirectional'.

3.4.9. *Access channel and rate*

This attribute describes the access channel type and the bit-rate of this channel. The access to an ATM link is only limited in terms of bandwidth by the available access-link capacity. However, only certain bit-rates will be allowed (i.e. there will be *granularity*), because this will lead to easier control of resources (connection admission control and usage parameter control). Two sub-attributes, describing access channel and rate, separately for user information and for signalling, have to be introduced for ATM.

3.4.10. *Access protocols*

These attributes characterise the protocol on the signalling and user-information channel at a given network termination or reference point. Seven sub-attributes have to be introduced to take into account the needs of ATM:

- Signalling access protocol - physical layer
- Signalling access protocol - ATM layer
- Signalling access protocol - AAL
- Signalling access protocol layer 3 (above AAL)
- Information access protocol - physical layer
- Information access protocol - ATM layer
- Information access protocol - ATM adaptation layer

3.4.11. *Supplementary services*

This attribute refers to the supplementary services associated with a given telecommunication service and is still being studied.

3.4.12. *Grade of service for the set-up and clear-down of VCC*

These attributes describe the network response time (specified in terms of the passing of specified layer-3 messages between specified reference points) and the probability of the blocking of a call attempt in the network (end-to-end blocking).

3.4.13.   *Performance of an established connection*

The performance is another area that is still under study within standards bodies at the time of writing this book, but examples of the types of parameter that could be included are listed below. When considering values for parameters, it must be remembered that the performance of a connection will depend on the traffic load; so values have to be defined for reference conditions.

Possible parameters:

- Cell-transfer delay: This is the total end-to-end transmission time of an ATM cell through the network, from the originating terminal equipment to the destination terminal equipment. This transmission delay includes the accumulated processing times in the switching stages, the accumulated transfer times over the links and trunks and the accumulated waiting times in the queues at switching stages. Part of this time is fixed for all cells in this connection and part is variable.

- Cell-loss probability: Cell loss may occur in an ATM network because of overflows in a cell buffer or because of uncorrectable header errors in transmission links. Because of the statistical behaviour of ATM traffic, cell loss cannot be completely avoided in the network. The cell-loss probability is defined as the probability that a cell is discarded in an end-to-end connection. This can be influenced by the value of the loss-priority bit. It can also be service dependent, because the statistical nature of the source (particularly its burstiness) will affect the probability of a cell being discarded by the network. Different services will require different cell-loss probabilities to be achieved.

- Cell-insertion rate: This is the number of cells in a certain time interval which are inserted (from other connections) in a connection.

- Bit-error probability in the information field.

The requirements on the values of these parameters to be achieved in an ATM network need to be defined assuming that certain standardised service mixes exist on the network.

3.4.14.   *Interworking possibilities*

This attribute describes the interworking between different bearer services and is yet another area still to be defined.

3.5. Mapping of services onto attributes

Having considered the range of services that might be available and the properties of the bearer services over which they are carried, it is then possible to derive a mapping of services onto bearer service attributes to determine the number of *different* bearer services that will be required. This area of work is still under study, but the principles can be seen from the examples in Table 3.2. As it is only an illustration, this table does not try to list all the combinations for a particular service. For instance, video-telephony could have an attribute of point-to-multipoint for conference calls and this is not shown in the table.

Table 3.2 Examples of mapping services onto bearer service attributes

Service & main features	Applications	Possible bearer-service attributes
Video-telephony		
• conversational service	• business conversations	• bi-directional symmetric
• moving pictures & sound	• social conversations	• point-to-point
	• tele-shopping	• on-demand
	• tele-education	
Video-mail		
• messaging service	• electronic mailbox	• on-demand
• moving pictures & sound		• point-to-point
		• unidirectional
Video-library		
• retrieval service	• entertainment	• on demand
• text, data, sound	• distance learning	• point-to-point
• still & moving pictures		• bi-directional asymmetric
Television distribution		
• distribution service	• entertainment	• permanent
• no presentation control		• broadcast
• moving pictures & sound		• unidirectional

Chapter 4

Networks and Network Elements

4.1. Network structure and elements

The main purpose of the ATM network is to provide ATM connectivity, that is an end-to-end ATM link for the transport of the user's information in ATM cells. To achieve this requires *switching* and *transmission*, together with the necessary *control* and *management*. This chapter considers the functions for switching, transmission and signalling (control); management of the network is treated in Chapter 11.

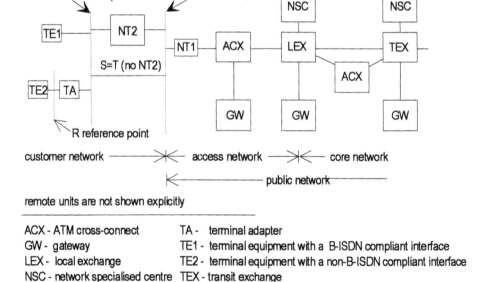

ACX - ATM cross-connect   TA -   terminal adapter
GW - gateway   TE1 - terminal equipment with a B-ISDN compliant interface
LEX - local exchange   TE2 - terminal equipment with a non-B-ISDN compliant interface
NSC - network specialised centre   TEX - transit exchange
NT - network termination

Figure 4.1 Generalised network structure

In the broadband network there are two separate network areas: the *customer equipment* (sometimes called the 'customer premises network') and the *public network*. The boundary between these two areas is the T reference point. The general network structure (Figure 4.1) is basically the same as that for the N-ISDN [8, 61] except for ATM-specific items (such as the ATM cross-connect). For simplicity, the figure does not show all possible network elements. It must also be emphasised that not all may be present in a particular scenario. Moreover, the figure does not show *remote units* (which may be multiplexers, concentrators or switches) since it is possible to think of the local exchange and remote unit combination as a distributed local exchange and this figure is not intended to show any geographical considerations. The ATM cross-connect function implies VP switching under network control, whereas the local exchange and transit exchange (also known as 'trunk exchange') functions will have the capability of on-demand call-by-call switching.

Terminal equipment and adapters are connected to the public network through the NT1 functional block. Within some customer premises networks there may also be an NT2 that performs PBX-like functions, ranging from the sort of features found on advanced PBXs to a simple system, with no internal switching, that provides a connection to the NT1 for a few terminals. The *functional block* NT1 is separate from the NT2 and provides a minimal set of functions for connection of the customer's equipment to the public network. The T reference point on the customer side of the NT1 is the usual (outside of the USA) demarcation point between the public and private networks.

The *implementation* of the NT1 and the NT2 must be separate in some countries for regulatory reasons; however, it may be technically and economically more suitable to combine the NT1 and NT2 in the same implementation.

On the public network side, NT1s may be connected directly to local exchanges (LEXs) or via remote units. Remote units may also provide switching (or cross-connection) without the need for switching in the LEX (i.e. the remote units are not under the control of the LEX). LEXs may be directly interconnected or connected through transit exchanges (TEXs): these TEXs may have call-by-call switching capability or may simply be ATM cross-connects.

Gateways provide a means of connection to other networks. In principle, they may be implemented anywhere in the network, at the remote unit LEXs, TEXs or on the customer premises, or indeed in some combination of these. The whole issue of gateways and interworking is considered later. It is worth noting here, however, that gateways will be used for the connections between the ATM network and public metropolitan area networks (MANs), N-ISDN networks and non-ISDN networks. However, in the customer premises network, non-ATM equipment (for example terminal equipment, LANs, private MANs) will be connected via blocks equivalent to a *terminal adapter* (TA), i.e. at the R reference point. There is perhaps some merit in referring to such blocks as 'LAN adapters' when they are used to connect to non-ATM LANs. This is illustrated in Figure 4.2.

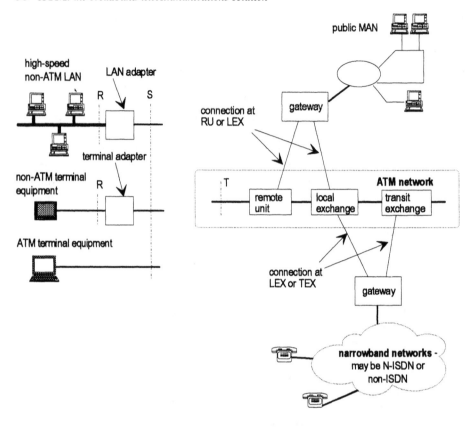

Figure 4.2 Connections between an ATM network and non-ATM networks and equipment

4.2. Interfaces

4.2.1.   Basic characteristics at S and T reference points

Figure 4.1 showed the references points S, T and R in the customer's network. Strictly, the letters S,T and R refer to reference points. Thus, when considering the characteristics of the interfaces at these reference points, the correct terminology is to talk about (for instance) the *interface at the T reference point*; however, for simplicity the term *T interface* is commonly used. To emphasise the broadband nature, the suffix B is often added to give T_B *interface*.

The standardised interfaces are those at the reference points S and T. These will be able to support simultaneously many combinations of services requiring different bit-rates (both CBR and VBR) including broadband and existing N-ISDN services, with the limit on bandwidth being the payload capacity of the appropriate interface. Interfaces at the R reference point may or may not have broadband capabilities.

In a simple B-ISDN implementation, there may be no NT2 function (equivalent to no PBX) and the S and T reference points are then exactly the same. Under this condition, the interface specification at the T reference point ensures that a simple broadband terminal may be connected directly to the T interface. However, at the time of writing, the characteristics of the interface at the S reference point for all types of configuration have not been finalised within the CCITT.

As explained in Chapter 2, there are two values currently envisaged for the gross bit-rate for the T_B interface, which may be optical or electrical:

- 155.52 Mbit/s: this interface is bi-directional with the same bit-rate in both directions
- 622.08 Mbit/s: two possible interfaces have been identified, the first having a bit-rate of 622.08 Mbit/s in one direction and 155.52 Mbit/s in the other and the second being a symmetrical interface at 622.08 Mbit/s.

It is worth emphasising again that these bit-rates are not available for user information, as illustrated in Table 2.5. This table showed that the maximum service bit-rate that can be supported on this interface is 135.631 Mbit/s, but it may be reduced still further by factors such as service delay and buffering requirements, the need to transport signalling and OAM cells and by the overheads of the AAL.

In addition to these currently standardised interfaces, there is a demand for a lower bit-rate interface (perhaps around 34 Mbit/s) provided over copper. The demand for such an interface comes from the world of data communications, where it is envisaged that existing copper plant in the customer's network can be re-used. However, such an interface may also be welcomed as an evolutionary step by network operators who would see it as a means of providing a relatively low-cost access for residential and small business customers, perhaps in a configuration that used fibre to a kerbside-cabinet with the final drop being copper (known as 'fibre to the kerb').

4.2.2. *Interfaces for distribution services*

The bandwidth of the user access link is large enough to support distribution (also known as 'distributive') services (video and audio). However, the network provider will wish to inject these services into the network as close as possible to the users to avoid using a large amount of resources by carrying this distribution traffic through the main network. These services will, therefore, be injected into the local part of the network; this may be at the local exchanges, the remote units within the local network or possibly the multiplexers in the access link. Two interfaces are defined for that purpose: one ATM-based and one non-ATM.

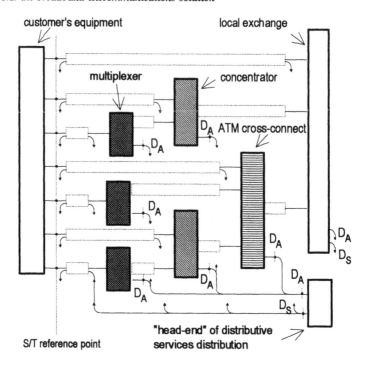

Figure 4.3 Possible injection of distribution services

The D interface is one possible interface implementation that has been proposed by one particular RACE project working in this area; although it has not yet been adopted as an international standard, it illustrates the principle. Figure 4.3 shows possible locations of the D interface which is an asymmetrical interface with two variants:

- D_A: an ATM interface for distribution services, which can be connected to every ATM node in the local network
- D_S: a non-ATM interface for distribution services, which must be connected to the access section, sharing only the transmission medium with other services

4.3. Customer equipment

As the interface at the T reference point is specified as a point-to-point interface, an NT2 function will be required in those applications where more than one terminal is required to be connected to a single access line: this is likely to be the vast majority of cases, especially for business customers. In some cases, the functionality of the NT2 will be low (for instance for residential customers). For others, it will be much greater, with the NT2 providing local switching. Here, the

NT2 may be an ATM PBX or it may be an ATM LAN, depending on the equipment available and the particular applications required by the customer.

Because the NT2 will be multiplexing the ATM traffic from the terminals, it will affect the traffic profile. Hence, the design of the NT2 must be such that it does not cause a legitimate traffic profile from a terminal to be distorted in such a way that the traffic then violates the parameters declared when the connection was set-up. Expressed another way, the cell-delay variation introduced by the NT2 must be within specified limits.

As broadband services and equipment are introduced, customers will expect some technical continuity in their own equipment. For instance, they may require that terminals used at present should be capable of being used, at least in the early stages, on new customer networks. However, the introduction of new applications over the new services will encourage them to buy new equipment to take advantage of the new applications.

4.3.1.   Business customers

At present, business networks usually provide one type of service, the dominant networks being PBXs for telephony/voice services and LANs for data services.

PBXs are increasingly digital, extending the capabilities of the N-ISDN to the customer's network. However, the 64 kbit/s circuit-switched capability has not proved attractive for data services within a customer's premises because of the relatively low bit-rate; faster LANs (with Ethernet and token ring) are now widespread, with even faster networks (FDDI and DQDB) appearing. Any change in the future will build from this position.

Initially, it is expected that LANs will evolve towards higher-speed networks and ATM LANs may well be one of the first applications of ATM. Such networks will allow new broadband services to be carried on the customer's site and this is likely to lead to greater demand for broadband connectivity between sites. For different sites of the same organisation, this may be achieved by an ATM cross-connect network, but more general interconnection between different organisations will require call-by-call switching.

New services available on high-speed LANs will include multi-media services, and one of the components of such services will be voice. The introduction of voice as such a component will be a major factor in the integration of voice and data services.

PBXs are also likely to evolve to support video services. Eventually, the PBX and LAN networks will be integrated in a single ATM network; before that, some form of interworking unit may be necessary.

4.3.2.   Residential customers

Residential customers are likely to be particularly sensitive to the cost of providing new equipment and such reluctance to spend money may hold back the penetration of B-ISDN services to residential customers. However, low-cost NT2 equipment may be introduced to include such functions as audio and video distribution and switching throughout the home, or switching control signals for home applications. In addition, new applications, such as tele-working, will lead to more business equipment being used in the home, with much or all of the costs being paid by the employer.

4.4. Access network

4.4.1.   NT1

In the customer premises network, the functionality is divided between the NT1 and NT2 in such a way that the NT1 (which terminates the public network and is maintained by the public operator) only contains the minimal set of functions required by all customers. An NT1 is required for the termination of every customer line and a universal low-cost implementation is therefore highly desirable. Those customers who wish to have more than one access to the public network will need a separate NT1 for each access connection.

The NT1 will include only those functions necessary to perform the physical termination of the access link to the customer (including some basic maintenance and looping capabilities) and those that are required by all customers. Because of its minimal set of functions, there should be only one design of NT1. This will be a low-cost unit capable of being used in all environments. Consequently, the NT1 will provide a single point-to-point T-interface and will not provide access to more than one NT2.

Transparency with respect to signalling with the public network is a requirement for the NT1, but the NT1 will include functions relating to the generation and reception of maintenance cells. Also, when the NT1 terminates *cell-based* operation and maintenance flows, cell delineation will be required.

The means of powering the NT1 is not yet defined. Both the NT1 and NT2 will require a minimum service (e.g. telephony only) to be defined and guaranteed during emergency conditions like power-down. These requirements are likely to be affected by regulatory constraints.

4.4.2.   Access link

The ATM access link, with its associated electronics, is likely to contribute a significant proportion of the cost of implementing an ATM network [1]. The most straightforward implementation of an access link is to use individual fibres from the local exchange to the customer, with possibly duplicated transmission for security in some cases. However, this is also the most expensive option.

To decrease the cost of the access network, remote units can be used. These will have the effect of shortening the final drops to customers and splitting the access into two parts: *access links* from the customer to the remote unit and *primary links* between the remote unit and local exchange. Naturally, there will also be cases where remote units are not used and the access links are connected directly to the local exchange.

For residential and small business customers, it will be essential to provide a low-cost access link. If a lower-bit-rate access is standardised then it would in some cases be possible to re-use existing copper connections, for instance coaxial cable CATV links. As explained later in Chapter 6, a very attractive route for low-cost access is the use of *passive optical networks* (PONs) which share the access link between a number of customers. Other factors, such as geography and the provision of existing ducts, will have a bearing on a real installation, but the principle of using concentration and sharing remains for those users with relatively small demand.

4.5. Switching nodes

4.5.1. Switching functions

Switching defines how a transmission path is routed through a network from the source terminal to the destination terminal, and how channels of intermediate links are associated with each other to form a connection between two end points. As explained earlier, the ATM transport network is structured as two layers, the ATM layer and the physical layer. The transport functions of the ATM layer are independent of the physical-layer implementation and are subdivided into two levels: the VC (virtual channel) level and the VP (virtual path) level.

Switching in an ATM network is very different from that in the N-ISDN. Although a connection is established at call set-up, it does not consist of a fixed bandwidth path that is exclusively available for the particular connection. Instead, cells from many connections are multiplexed onto a link, then queued at the switching node before each cell is individually switched to the appropriate destination link.

The bulk of the traffic in the network will be generated by the business customer, with a significant proportion being based on semi-permanent connections or leased lines. Managing this share of the traffic in an optimal manner results in significant cost savings and high revenue: such traffic management is possible with the use of flexible and re-configurable ATM cross-connects in the network.

In general, a connection in an ATM network can be established *on-demand* or the connection can be *semi-permanent* or *permanent*. With on-demand switching, the establishment and release of a connection is done by using signalling procedures in the *control-plane* of the protocol reference model. Semi-permanent connections between agreed points may be provided for an indefinite period of time after subscription, for a fixed period or for agreed periods during a day, week or other interval. Permanent connections are available to the customer at any time during the period of subscription between the fixed points requested by the customer when the permanent connection was ordered. Both semi-permanent and permanent connections are set up by the management plane.

In a more formal way, it can be said that in control plane communication a user manages (establishes, releases and maintains) a VPC/VCC by sending control-plane messages through a signalling VCC that is terminated at a VC-switch. In management-plane communication, the connections are established, released and maintained by the VP or VC cross-connect using the network-management function.

There are three types of *switch* defined in an ATM network and, as explained above, they are directed by control plane functions; the equivalent *cross-connects* are directed by management-plane functions. A *VP switch* is a network element that connects VP links and it translates VPI values between input and output; a *VC switch* is a network element that connects VC links, translating VCI values, and terminates VPCs; a VP-VC switch is a network element that acts both as a VP switch and as a VC switch.

4.5.2. Remote unit

The remote unit is an optional network element (dependent on the requirements of a particular access network) situated between the NT1 and the local exchange. It is clear that several local network topologies could be used in a broadband ATM network. These depend on local circumstances, such as subscriber density, the ratio between business and residential customers, the type of services and so on.

Remote units can, in principle, perform multiplexing, concentration or local switching, as illustrated in Figure 4.4. A topology containing remote units is particularly suitable for situations where the traffic from individual customers is insufficient to justify direct fibres between the customer and local exchange. In this situation, the remote unit performs *concentration* of the traffic, the degree of concentration achievable with ATM depending on the traffic characteristics. In remote areas, there may also be a case for the remote unit performing *local switching* to avoid having to transport traffic between tributaries of the same remote unit to the local exchange and back.

Architecturally the remote unit is part of the local exchange, because no interface is defined between a remote unit and a local exchange. However, it is common usage to treat the remote unit as a separate network element, particularly when considering the geographical aspects of an installation.

As a functional block, the remote unit will implement some or all (depending upon whether switching is required) of the following functions:

- Concentration of traffic
- Terminating the external links
- Usage parameter control
- Connection admission control
- Switching

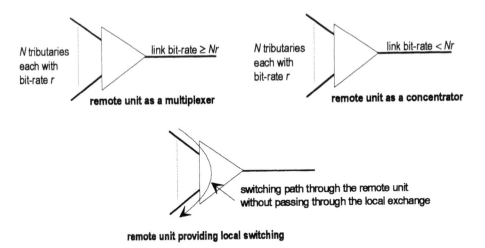

Figure 4.4 Illustration of a remote unit providing multiplexing, concentration or local switching

The presence of distribution services in the local network affects remote units. In order to minimise the network capacity required, it is advantageous to insert distribution services into the network at the point nearest the customer: this is the remote unit (if a remote unit is used in that connection). For this application, the remote unit requires the capacity to handle these services.

4.5.3. Local exchange

The local exchange (LEX) provides interfaces on the customer side with NT1s and remote units; on the network side it connects to transit exchanges and other local exchanges. The local exchange is able to perform VCI/VPI routeing on all its external links.

As a functional block, the local exchange will implement the following functions:

- Switching
- Terminating the external links
- Usage parameter control
- Call control
- Overload control and throttling
- Charging functions
- Routeing functions
- Control of remote units attached to the local exchange

Distribution services have an implication for local exchanges as well as for remote units. Where remote units are not used, the local exchange is the nearest public-network function to the customer and must distribute these services. For this application, local exchanges require a broadcast capability.

4.5.4. Transit exchange

The transit exchange (TEX) is sometimes called the 'trunk exchange'. It interfaces only to local exchanges and other transit exchanges. The functional differences between a local exchange and a transit exchange are as follows:

- The transit exchange has to provide termination functions on the network side only, so it has a reduced set of interfaces
- The transit exchange has to provide large switching capabilities without concentration so it will have a larger switching network
- Since the traffic will already have passed through UPC functions, usage parameter control as such will not be required in a transit exchange
- The transit exchange does not need a charging capability

If distributive services are being transmitted across the network (rather than directly from the service provider to each local exchange, for example by satellite), transit exchanges also require broadcast capability.

4.6. Transmission

ATM is a transfer mode; in principle it is independent of the transmission system provided that the required bit-rate can be supported. In the core network, the existing plesiochronous digital hierarchy (PDH) transmission and synchronous digital hierarchy (SDH) transmission will be used to carry ATM cells. In the access network, SDH transmission and cell-based transmission will be used.

When ATM is introduced, there will still be many transmission links using the PDH hierarchy as well as others using SDH. Hence, at the outset, it will not be possible to carry ATM using only SDH; PDH transmission adapted to ATM will have to be used. This is likely to be only an interim step until there is national SDH coverage and there will be an evolution of the transmission network from PDH to SDH, with possibly some parts using a cell-based transmission hierarchy.

It is important to emphasise that the introduction of ATM does not require the existence of an SDH transmission network.

4.6.1. PDH transmission

Transmission systems to existing G.702-G.703 standards will still be in use when ATM is introduced, so it is expected that they will have to support the transmission of ATM in the core network (at bit-rates of 140 Mbit/s upwards). Use of the PDH offers the following advantages:

- New line equipment is not required (existing multiplexing equipment and line plant retained)
- It provides immediate national coverage
- It could initially carry ATM on some tributaries and existing services on others

However, there are also disadvantages:

- It can only be an interim solution
- Operation and maintenance not as easy as with other options
- A mapping of ATM cells to the PDH structure is required
- It has poor flexibility for network reconfiguration.

4.6.2. SDH transmission

New SDH equipment is currently being introduced into public networks and this equipment will be used to carry ATM traffic in both the core and the access network. The only major disadvantage of SDH transmission is that it is not sufficiently widespread at the introduction of ATM. Otherwise it offers significant advantages:

- New equipment is not required if the SDH system is installed already
- It could be the target for transmission
- Operation and maintenance is easy
- It has good flexibility for network reconfiguration
- Although a mapping of ATM cells to the SDH structure is required, this has been defined by standards bodies and ATM-in-SDH is a standard physical-layer interface with chip sets already available.

4.6.3. Cell-based transmission

There has been some initial work on defining a new cell-based transmission system for ATM, the current assumption being that it will only be used in the access network. Such a transmission system would offer similar advantages to SDH with good flexibility and easy operation and maintenance. In addition, there would be no need to map ATM cells to a physical-layer frame structure.

4.7. Signalling

4.7.1.   *Requirements for B-ISDN signalling*

The signalling protocol used in the ATM-based B-ISDN [5] must support different service types. Future broadband services will include features such as mobility, interaction with mobile services, the intelligent network concept and new facilities for network management. ATM supports the logical separation of signalling from user information and it has the flexibility to support those services that require multi-connection, multi-party and/or multi-media calls.

Requirements for B-ISDN signalling fall into three groups:

(i) Capabilities to control ATM virtual channel connections and virtual path connections for information transfer

- Establishment, maintenance and release of ATM VCCs and VPCs for information transfer; the establishment can be on-demand, semi-permanent or permanent, and should comply with the requested connection characteristics (e.g. bandwidth, quality of service)
- Support of communication configurations on a point-to-point, multipoint and broadcast basis
- Negotiation of the traffic characteristics of a connection at connection establishment
- Ability to re-negotiate source traffic characteristics of an established connection.

(ii) Capability to support simple, multi-party and multi-connection calls

- Support of symmetric and asymmetric simple calls; asymmetric calls have a different bandwidth in each direction
- Simultaneous establishment and removal of multiple connections associated with a call
- Addition and removal of connection to and from an existing call
- Addition and removal of a party to and from a multi-party call
- Capability to correlate on request connections composing a multi-connection call
- Reconfiguration of a multi-party call including an existing call, or splitting the original multi-party call into more calls.

(iii) Other capabilities

- Capability to re-configure an already established connection, for instance, to pass through some intermediate processing entity such as a conference bridge
- Support of interworking between different coding schemes
- Support of interworking with non B-ISDN services

- Support of failure indication and automatic switching for semi-permanent and permanent connections.

4.7.2. *Signalling evolution*

B-ISDN signalling will evolve towards a target from an initial release called 'Release 1'. This is based on existing signalling [74] and at the time of writing this book it has reached the following position:

At the UNI

- It is based on Q.931
- A subset of layer-3 messages has already been defined
- The first description of new and modified information elements (ATM Traffic Descriptor) is available
- A new protocol discriminator to identify B-ISDN messages has already been defined
- New procedures for call set-up and call-release (simultaneous call/bearer control) will be included.

B-ISDN layer-3 signalling at the NNI interface

- It is based on Signalling System No 7
- It uses a new B-ISDN ISUP

The target signalling system will be developed in several steps, a baseline document with harmonised signalling requirements already being available. The major difference between the target and Release 1 is the separation between call and connection, the control plane in the protocol reference model being divided into *call control* and *connection (bearer) control*. At present, there is no decision as to whether there will be a common protocol for UNI and NNI, or whether the new protocol will be based on an existing one.

The separation of call control and bearer control (Figure 4.5) offers several advantages:

- Not all nodes on the network need to be equipped with call control; intermediate switches can be equipped with bearer control only.
- Multi-media services can be controlled in an efficient and flexible way: within an existing call, connections can be added or released without releasing the call.
- Multi-party calls can be controlled in an efficient and flexible way: parties can be added and excluded without releasing the call; multi-party calls can be re-configured.
- Establishment and release of connections and calls can be done simultaneously and sequentially; the latter offers the possibility of negotiating call acceptance before the connection has been set up.
- There is the possibility of database access without connection establishment.

- Charging and tariffing can be dealt with more easily, because it can be based on call procedures and connection parameters.
- It leads to easier introduction of new services.

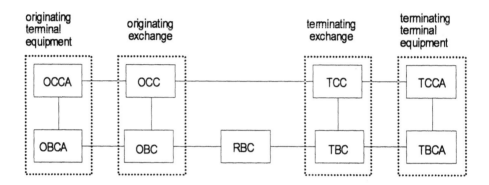

OCCA: originating call control agent
OCC originating call control
TCC: terminating call control
TCCA: terminating call control agent

OBCA: originating bearer control agent
OBC: originaring bearer control
TBC: terminating bearer control
TBCA: terminating bearer control agent
RBC: relay bearer control

Figure 4.5 Separation of call control and bearer control in target B-ISDN signalling

4.7.3.   Call control and bearer control functions

Call control functions are:

- Identification of a requested telecommunications service
- Supervision of negotiation and of the reservation of virtual channels and network resources
- Choosing of the next call relay
- Supervision of the connection-control link-by-link procedures
- In a call offering, call control indicates the characteristics of the service
- Control of supplementary services

Bearer control (connection control) functions are:

- Setting up of a connection (with or without knowledge of call control)
- Negotiation and reservation of virtual channels and network resources
- Choosing of the path
- Bearer control is a link-by-link procedure
- In a call offering, bearer control indicates when a connection is established

Quality-of-service negotiation is a task for call control, not bearer control. This is because, at the call negotiation phase, it should already be known if the required quality of service can be provided.

Call control includes a function 'negotiation of call structure', since it might be desirable to negotiate how the set of connections should be composed in order to support the service requested. Call control includes a function 'service request acceptance' to indicate that the user interacts with the call control and not with the bearer control.

Chapter 5

Network Performance

5.1. Introduction

There are several performance measures that are specific for ATM, such as cell-transfer delay and delay variation, cell-loss probability and cell-insertion rate. Cell-delay variation and cell loss arise from queuing and buffer overflow, respectively. They occur because of the characteristics and statistical variation of ATM traffic; they are therefore important measures of performance in ATM traffic engineering. ATM-layer phenomena depend on both ATM-layer and physical-layer performance. Physical-layer effects are not a subject of ATM traffic engineering *per se*. However, because random bit errors can cause cell loss (for example), it is important to be able to distinguish the reasons for cell losses when this phenomenon is measured on a real system.

Thus, it is instructive to address the causes and effects of network performance in *layers* [82, 98]. The physical-layer performance is independent of the performance of the ATM layer; however, the latter is affected by the former. The performance of both layers will affect the performance of the AAL. Aspects of all three layers have an effect on the quality of service experienced by the user. This interdependence of cause and effect between layers is illustrated in Figure 5.1.

5.2. Delay

5.2.1. Components of delay

The overall end-to-end delay experienced by signals as they traverse the network from source terminal equipment to destination terminal equipment arises from the transmission length and the delay in different network entities and includes cell assembly/disassembly delay at the ATM network edge. The delay in the customer's network is also included in this definition. Multiple cell assembly and disassembly processes are obviously to be avoided if the delay values are to be minimised.

The delay through an ATM switching element is expected to be negligible compared with other delays, although this may vary with the traffic load on the switch due to the queuing delays.

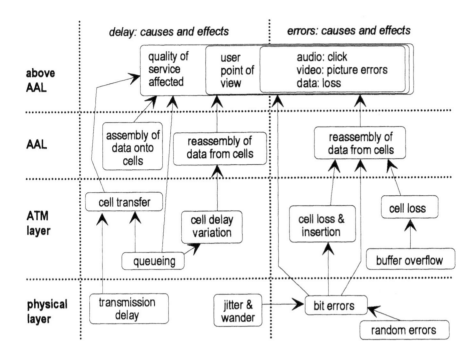

Figure 5.1 Delay/error effects, and propagation through the layers

Delay should be minimised wherever possible. The measures that can be taken include the following:

- Partially filling the cell, which would lead to a loss of transmission efficiency
- Avoidance of an excessive number of multiplexing stages in the transmission by providing direct high-bit-rate links whenever possible
- Minimising the number of cell assembly/disassembly points when interworking by avoiding, wherever possible, passing ATM traffic through transit, non-ATM networks
- Avoidance of an excessive number of intermediate transit switching points by using as many direct routes as possible and reducing the number of levels in the switching hierarchy
- Minimising the delay in the customer's equipment, at both ends of the connection.

5.2.2.   Delay in the physical layer

Jitter is defined as the short-term variations of the significant instants of a digital signal from their ideal positions in time. It is well understood and well controlled in existing digital networks, and can be reduced in magnitude by the use of jitter reducers. *Wander* is defined as the long term variations of the significant instants of a digital signal from their ideal position in time. It can arise as a result of changes in the propagation delay of transmission media and equipment. Slips as the result of wander will be rare provided equipment input ports can tolerate wander in accordance with the requirements of [19, 20]. Both jitter and wander can result in bit errors in the physical layer, which in turn cause errors in the layers above. *Transmission delay* results from the length of a connection, and for optical fibre cable systems is 6 μs/km.

5.2.3.   Delay in the ATM layer

Cell-transfer delay is a constituent of the overall end-to-end delay but it *excludes* the cell assembly/disassembly time (a function of the AAL) and the delay in the customer's network. Cell-transfer delay is therefore confined to the delay caused by the transmission length (i.e. due to the physical layer) and that arising in different network entities (such as switching nodes). It varies according to the instantaneous loading of the network, and this has two consequences: cell delay variation and differential delay between service components for multi-component services. Control of the network load is important, to maintain these effects within limits. *Cell delay variation* is the variable component of cell-transfer delay. It is caused mainly by variations in cell delay in the network nodes. Its causes include queues in ATM switching fabrics (call-by-call and cross-connects), queues in line equipment used for medium access and queues within receivers for adaptation of cell-rates.

5.2.4.   Delay in the AAL

A user source, such as digitised speech at 64 kbit/s, requires conversion in the adaptation layer into the ATM cell format. This process (*cell assembly*) introduces delay, since the cell is not normally released until the information field of the cell has been filled. For example, a 48-octet information field carrying 64 kbit/s speech would, if filled, incur a cell-assembly delay of 6 ms. Since the customer's equipment includes the functions of cell assembly/disassembly and switching, there is clearly a need to assign a portion of the permitted overall delay for delay-sensitive services (such as voice telephony) to the customer's own network. An allocation based on the standards for the existing narrowband networks would normally allocate a maximum value of 5 ms for the delay the customer's equipment at each end of the connection and this is clearly not achievable for 64 kbit/s sources in an ATM network.

The ATM cells arrive with no synchronisation with the destination. Certain services (e.g. constant-bit-rate services) require end-to-end timing functions and the receiving AAL protocol entity must reconstruct a constant or variable bit-rate stream from a sequence of cells that arrive with varying delay. A receiver buffer compensates for the cell delay variation, but slight deviations between transmitter and receiver clocks can cause overflow or depletion of this buffer. The user is unlikely to notice any degradation if the receiver slightly adapts its rate of emptying the buffer so that overflow and depletion are prevented.

5.2.5. Consequences of delay

5.2.5.1. Differential delay between VCCs
The difference in delay of two real-time VCs between the same two network endpoints may be a problem for multi-connection calls (e.g. disturbances to lip and speech synchronisation in video telephony, if the *differential delay* is too great).

5.2.5.2. Echo effects
The total delay can cause *echo* problems for voice transmission (especially when interworking with the PSTN), unless precautions are taken [58]. The main cause of echo is balance-impedance mismatch at 4-wire/2-wire terminations. A mismatch at the far end of a connection causes a reflection to return to the near end. Since an ISDN uses 4-wire transmission throughout, this occurs only at customers' terminals and the minimum return loss is 12dB. Handsets can also cause echo because of acoustic coupling between earpiece and mouthpiece, but the attenuation of this is normally greater than 36dB.

The subjective effect of echo depends on the attenuation of the transmission path. For a digital network this is zero. The overall loudness rating (OLR) of an ISDN connection is determined solely by the sensitivities of the telephones and the parameters of the speech codecs. For digital telephones, this is targeted at 10dB. Consequently, the echo-path loudness rating is only 10dB more than the balance return loss.

The annoying effect of echo increases with the delay. Permissible values, based on subjective tests are available, given the value of one-way delay. In full ATM networks, it is expected that the echo effects generated by the digital telephone set will be within the requirements.

For the interworking of ATM with other networks, the echo caused by the imbalance of 2-wire interfaces will introduce discomfort to most long-distance calls and cannot be neglected. This will require an echo-control device. For short-distance calls, this discomfort does not exist provided that the cell assembly / disassembly delay is kept small. For international calls, interworking with echo cancellers provided by international gateways must be taken into account.

Examples of the delay requirements for existing networks are taken from the UK PSTN and from the planning rules applied by Telekom to the German national network.

In the UK PSTN, the one-way delay for 95% of telephone connections should not exceed 23 ms, except for international calls where the one-way delay to or from an international gateway should not exceed 23 ms. For national calls the allowance of 23 ms comprises:

- 2 ×5 ms for the two sets of customer's equipment
- 2 × 3 ms for the two local ends
- 7 ms for the transit network

Since the cell assembly/disassembly time of a fully filled ATM cell is 6 ms at 64 kbit/s it is clear that the delay requirements for the customer's equipment cannot be met.

In the German network similar delays are specified, the overall requirement being that if the end-to-end transmission delay exceeds 25 ms then echo cancellers will be fitted.

As explained earlier, the two solutions to overcome problems caused by delay are:

- To accept the delay and introduce echo cancellers
- To use partial filling of cells to reduce the delay

At first sight, echo cancellers appear to be an expensive solution. However, as they become widespread for other purposes (for example in mobile networks) they will become much cheaper and are likely to provide the preferred solution.

5.2.6.   Errors

In ATM networks, different types of errors could happen from a layer viewpoint. These disturbances affect the ATM network performance by causing cell loss, cell insertion, loss of cell delineation and corrupted data. Transmission errors do not affect delay and delay variations.

5.2.7.   Errors in the physical layer

The problems of synchronisation-loss and data affected by bit-errors are well known in existing transmission systems. In ATM networks, bit errors in the cell header cause the loss of cell delineation (and hence cell loss or cell insertion). Only a flow of valid cells crosses the boundary between the ATM and the physical layers. No error control is performed on the information field which is transported transparently by the ATM layer. Thus, errors in the information field of an ATM cell will affect the performance of the AAL and the layers above.

If proper cell delineation is lost, it will not be recovered for another 6 or 8 cells. This leads to the loss of the order of 300 bits. However, these cells may not all be from the same connection or indeed may not be carrying user traffic. The effect of loss of delineation on an individual connection is, therefore, difficult to quantify.

5.2.8. *Errors in the ATM layer*

The main cause of errors due to the ATM layer is buffer overflow and the main reason for this potential overflow is the stochastic variation of the traffic being queued. Traffic-control and resource-management functions (described later) must be implemented in the ATM layer to minimise the likelihood of such overflow.

5.2.9. *Errors in the AAL layer*

The information field in the cell transparently crosses the ATM layer without any error control. Since this field is also the AAL protocol data unit, a physical-layer error in the information field will also be present in the AAL. The different AAL types provide different approaches to the handling of errors in the information field. For some applications the SAR and/or CS may be empty and any errors are, therefore, passed directly to the layer above the AAL. The functions that can be provided by the AAL include the following, but it must be emphasised that not all are available in every AAL type.

- Handling of, and compensating for, lost or misinserted cells (using sequence numbering)
- Monitoring of AAL protocol control information for bit errors
- Monitoring of user information field for bit errors and possibly taking corrective action
- Monitoring SAR service-data-unit sequence-integrity.

Chapter 6

Network Evolution

6.1. General technical aspects of network evolution

It is possible to introduce ATM first into the public network or into the private network and indeed there are already ATM products (such as ATM switches and interface cards for workstations) available for the customer's network. Private-network use of ATM may well be the first use of the technology, but the public network must evolve in parallel to provide the wider-area connectivity that customers will require.

There will be a range of alternative strategies or plans to introduce B-ISDN services into the public telecommunications network with the ultimate aim of evolving to a single universal network that will support all services [6, 49, 52, 56]. If the introduction is to be successful it must be well planned, taking into consideration the wide range of existing networks and services and their lifetimes as well as the effects on the end-users. The evolution of services is fundamental, as the only point of having the network is to provide services the customer is willing to buy.

The time scale of the introduction is important. It depends on the availability of new services, the supporting technology and products for those services and, most importantly, the demand for the services. The demand is, of course, strongly influenced by the tariff policy, particularly for residential and small-business customers. To minimise the increase in costs, both for the user and for the operator, it is essential that the integration strategy should begin with the integration of existing services, equipment and networks.

An important point is that different segments of a network can, to a limited extent, evolve separately. For instance, the access network in a particular geographical area could evolve completely to the target before the national network evolves to its target form: all services would be carried on the ATM access link, but narrowband services would be connected by gateways to existing networks. It is worth considering each segment in general terms before looking at specific details.

broadband connections   connections within existing networks

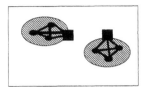 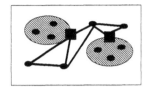

Within the ATM area all services (including narrowband) are carried on the ATM network.

a) substitution approach

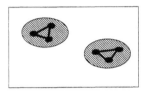 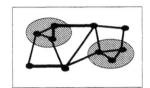

Within the ATM area only broadband services are carried on the ATM network and narrowband services are supplied by the existing network; there is no connection between ATM islands.

b) island approach

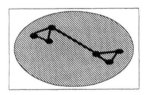 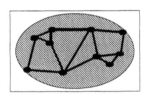

Within the ATM area only broadband services are carried on the ATM network and narrowband services are supplied by the existing network; the ATM area overlays a wide geographical area.

c) overlay approach

---

     geographical area where ATM is available

●   customer

■   gateway between broadband and narrowband networks

Figure 6.1 Substitution/island/overlay approaches

6.2. Substitution/island/overlay introduction approaches

There are three logical ways in which broadband networks could be introduced: *substitution, island* or *overlay*, as shown in Figure 6.1. In that figure, the lines representing connections are for illustrative purposes only and are not intended to represent connections between customers and exchanges.

It is worth emphasising here that each of these approaches represents only the first phase in the introduction of the ATM network: the target network is a fully-integrated broadband ATM network carrying all services throughout all countries.

The *substitution approach* consists of replacing all existing networks within a specific geographical area with a single ATM network which integrates existing narrowband services onto the broadband network. This approach needs a gateway in each ATM island to the surrounding existing networks.

The *island approach* consists of introducing an ATM network carrying broadband services to a specific geographical area although the existing narrowband services would be retained on the existing network. Using this approach broadband services could be introduced into the major financial districts of various cities throughout a country without disturbing their existing services. Although users within each district have broadband connectivity, those in differing islands have not and can only be connected by the existing narrowband networks via gateways.

The third approach is the *overlay network*. Here broadband services are offered to users throughout a whole region (perhaps a country) over and above the existing networks. This compares with the localised approach offered by the island strategies. Such an approach makes broadband services available to users over a much wider area and does provide wider broadband connectivity, but it may lead to long access links from the users to a possibly centralised switch.

The island type approaches tend to appeal to market-led operators operating in a liberalised environment, whereas the overlay strategy is more likely to be adopted by those operators who use an investment-led approach to their networks. The geography of a region and the nature of the businesses there are also important factors.

In practice it is likely that a combination of these approaches would be used. For instance, an overlay of ATM cross-connects to provide national coverage of 'leased line' services could be used at the outset, with islands of ATM call-by-call switching added as demand for new broadband services increases.

6.3. Customer groups

The concept of *customer groups* is a useful means of sub-dividing the customer base for network-planning purposes. The users of telecommunications services are generally split into two large categories of customers: *residential* and *business*, with the business category being further subdivided into *small business, medium business* and *large business*. Some authors introduce another category, the *professional user*, but this can generally be included within the small business category.

This division into customer categories is useful when considering the different types of access network (considered later) or for providing a mapping between types of customer and likely teleservices that customer would use.

Although customer groups are given names that broadly describe the type of customer being considered in that group (e.g. 'Medium Business'), there needs to be a more exact definition of the range. This is accomplished by the *measure descriptor* that describes the group in terms that can be readily visualised. Various parameters could be used to perform this function, for instance the number of employees or the number of calls made. The CCITT has used the number of employees at that site as the basis for the definition, but this measure does not take into account the type of business: for instance the headquarters of a financial organisation will have a greater demand for telecommunications services than an assembly plant, even if the number of employees on the site is the same. A better measure is the number of PSTN exchange lines. It should be noted that this is used only as a description of that group and that demand parameters are treated separately. For instance, a medium-business customer group may have a measure descriptor of between 11 and 100 exchange lines; the demand, in terms of narrowband PSTN connections, for that group will be a separate value within that range that represents the average number of connections in a particular scenario.

The *service demand* is a series of sets of values, one set for each customer group at any one time. Each set of values represents the average demand for each type of service for one particular group: the choice of services to make up the set and the average demand is a matter for the particular scenario being considered.

Population size is a parameter that takes on one value for each customer group at any one time, each value being the number of customers of that particular type in the scenario being considered. The numerical values chosen are a matter for the particular scenario being considered.

6.4. Fibre in the access network for business customers

Small businesses (which include the 'professional' user) are not included in this discussion, because the demand from each is likely to be relatively small and connections are more likely to be made using the same techniques as those for residential customers.

A fundamental requirement for the introduction of B-ISDN is the provision of a broadband access link to the customer. For large and medium business customers, it is envisaged that there would be a direct fibre connection between the customer and B-ISDN. For large businesses, this may be connected directly to the LEX, but for medium businesses it is perhaps more likely to be connected via a remote unit.

The following three sections describe different methods of provision of service, for existing services as well as for broadband, to the business customer once fibre has been provided. The three approaches are summarised in Figure 6.2 and Figure 6.3 shows how each approach may evolve to the target network [48].

It must be emphasised that each approach allows for the provision of all or any broadband service.

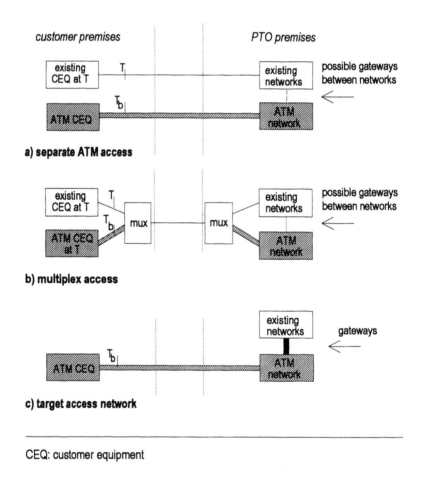

CEQ: customer equipment

Figure 6.2 Possible access configurations for business customers

6.4.1.   Separate access for broadband

In this approach B-ISDN services are offered to customers over a separate access link to the ATM local exchange (Figure 6.2a). There is no disturbance to existing services.

Advantages:

- Lowest cost start
- Potential early generation of revenue
- Minimum generation of new hardware
- No impact on narrowband access network
- Allows early introduction of end-to-end ATM connectivity

Disadvantages:

- Minimal integration between new and existing services
- No simplification of network operations and maintenance
- Once installed, customers may be reluctant to dispense with their narrowband equipment and move to full ATM access

6.4.2. Multiplex access system

This approach uses a single broadband access link to the customer carrying a multiplex of narrowband and ATM services. Before being presented to the customer, the composite traffic is de-multiplexed so that again the customer can retain existing narrowband equipment. Two possible techniques that could be used for the multiplex are *optical multiplex* (multiplexing at the *physical layer*) and *ATM multiplex* (multiplexing at the *ATM layer*).

With optical multiplexing the narrowband and broadband services would be carried on separate wavelengths on the access-link fibre. It could be considered that such a technique is very close to the separate access. If the de-multiplexing is carried out on the network operator's premises, the customer would not be able tell whether there was separate or multiplex access. With the ATM multiplex, all services would be carried by ATM and the narrowband traffic separated before presentation to the customer. The multiplex equipment would, in this case, have to perform the adaptation required to transport the narrowband services over ATM. This approach is nearer to the target than optical multiplexing as it may be possible to replace the de-multiplexing equipment at the exchange end with an integrated connection to an ATM exchange as a step on the evolutionary path.

Advantages:

- Moderate cost start
- Potential to generate new revenue quickly
- Early introduction of end-to-end ATM connectivity
- Begins to simplify local network operations and management

Disadvantage:

- For the ATM multiplex, the conversion to and from ATM for narrowband services could cause problems in terms of delay

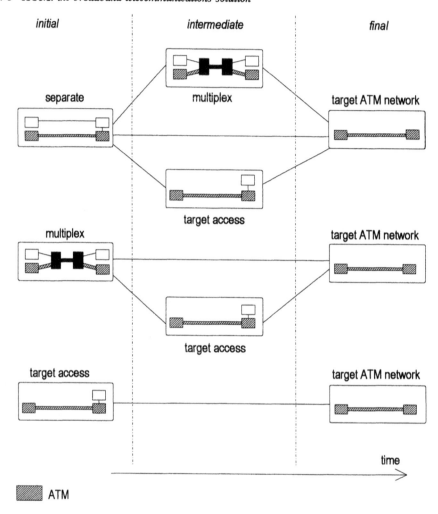

Figure 6.3 *Transition steps between access configurations*

6.4.3.   *Target access*

By the time the introduction of ATM networks is being considered, it is likely that there will be customer premises equipment using ATM technology. This approach aims to offer the customer the fullest use of ATM private equipment by providing full ATM access to an ATM exchange (Figure 6.2c). Existing narrowband services will be carried on this ATM access as well as new broadband services, access to the existing networks being provided by gateways between these networks and the ATM network.

This approach could be an evolution from other, less-comprehensive introduction strategies, or it could be an introduction approach in its own right.

Advantages:

- It is the target network
- Maximum potential for revenue
- Maximum flexibility for customer
- Greatest simplification in operation and maintenance for customer access

Disadvantages:

- High-cost start because of the need for gateways to non-ATM networks and interfaces (between non-ATM and ATM equipment) to be available
- The ATM network must immediately provide a quality of service to match existing services and must offer high standards of reliability

6.5. Residential customers and distribution traffic

The serving of residential customers provides both an opportunity and a challenge. The opportunity is to provide new services that will generate new traffic and hence new revenue. The challenge is to provide these services at an acceptable cost, bearing in mind that some of the services that can be provided over B-ISDN will be competing with services already available; for instance, video retrieval will be competing with the traditional video library, costing only a few ECU per night, or with satellite broadcasting.

Factors that will encourage the introduction of new higher-bit-rate, services to the residential customer are:

- Increasing prosperity leading to a higher disposable income
- Increase in the number of people using *tele-working*
- Reduction in costs from technological advances and from economies of scale once the fibre B-ISDN has been provided to business customers
- Political decisions to improve the telecommunications infrastructure for the residential customer

It is important to remember that the customer will see him/herself paying for services rather than the network; thus the key to the success of broadband residential services will be in devising attractive services that can be delivered at an acceptable cost.

A major component of the cost of the provision of broadband services to the residential customer is the cost of the access network, so it is important to devise economical methods of access. The discussion on the type of access method to be used must consider the type of service to be carried, in particular the bandwidth requirements of the service, which are summarised in Table 6.1. In this table, and indeed the rest of this book, the term 'PON' is taken to imply bandwidth sharing, as in the original form of passive optical networks for telephony. However, it is, of course, possible to use passive optical techniques independently of whether there is a sharing of bandwidth. The different methods of providing access are shown in Figure 6.4.

Table 6.1 Types of access

	Shared bandwidth	*Exclusive bandwidth*
Low bandwidth:	Fibre-to-the-kerb Passive optical network	
High bandwidth:		Direct fibre Logical star

Essentially, shared usage techniques lead to lower bandwidth availability than those where all the bandwidth is available to a particular customer.

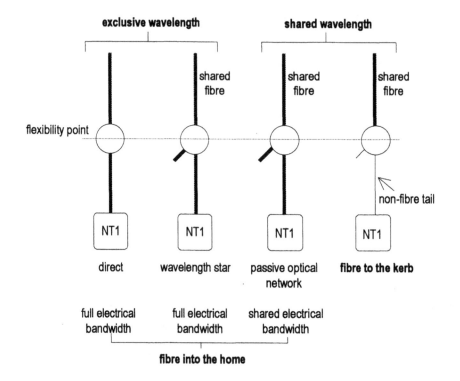

Figure 6.4 Possible access configurations for residential customers

The term 'fibre-to-the-home' is often used in the literature, but is not used here: the reason is that, apart from 'fibre-to-the-kerb', all the approaches bring the fibre to the customer's premises. The abbreviation 'F-T-T-C' is often used for 'fibre-to-the-kerb', being based on the American spelling: 'fiber to the curb'. In any case, there is not a great distinction between any of the access methods in terms of physical infrastructure. Each approach assumes that physical fibres will probably be laid in a tree structure with splicing of fibres to form larger cables the nearer one is to the head end. The major differences are:

* Whether the tail from the final distribution point (the 'kerb') is fibre or not
* Whether there is a shared bandwidth for interactive services, so requiring some form of protocol to manage the sharing
* The number of wavelengths used on the fibre
* Where the downstream gating filters are situated: final distribution point or in the NT1

6.5.1. *Fibre to the kerb*

Fibre-to-the-kerb has been proposed as a method for reducing the access cost, because optical-electrical conversion equipment is shared between a group of customers and the need to replace copper cables into the customer's premises is obviated. This technique would be particularly useful if a lower bit-rate ATM interface were to be standardised. An alternative using the same principle would be the use of radio tails from the cabinet to the home. Where allowed by the regulatory body, such an approach might be favoured for providing quick access in a competitive environment.

Advantages:

* There is less disruption to the customer if existing copper could be used or radio tails were acceptable.
* The number of electrical-optical conversions is minimised so reducing costs. One converter is required in the cabinet for each fibre and the lower-bandwidth copper tails allow one fibre to serve more than one customer.
* If the customer has CATV already, it might be possible to use the CATV cable into the house as the copper tail so reducing the installation costs.

Disadvantages:

* The bandwidth available to customers is fairly low and would be insufficient for some services, such as video retrieval.
* Provision of distribution TV services on the same access would be complicated unless using existing CATV installation.
* Once installed there might be resistance to change, so hindering the introduction of the full B-ISDN.
* It is envisaged as being only an initial stage.

6.5.2.   *Passive optical networks*

The key to the success of B-ISDN, and its associated services, with the residential customer is the provision of a fibre access system that carries the basic telephony service as well as entertainment (TV and radio). Attempts are already being made at using passive optical networks (PONs) as an economic access method for just telephony services. Telephony service on a passive optical network is generically called TPON (but not to be confused with the specific BT system also called 'TPON'). This scheme (Figure 6.5a) uses TDM and TDMA transmission methods over a shared optical fibre from the local exchange to the distribution point where optical power splitters fan out to individual customers. The important elements in the passive optical approach are:

- The tree structure, leading to sharing of fibre
- The use of passive, rather than active, splitters

Such networks may well be viable because they show savings on the cost of installing new copper pairs and on the cost of maintenance compared with the conventional star-configured copper cable network. They also provide a better transmission infrastructure.

The principles of the basic passive optical network can be extended to allow broadband distribution services to be carried. In this form, known as BPON, narrowband interactive services can be carried over the shared bandwidth of the passive optical network and distribution services carried on a separate wavelength using optical multiplexing techniques (Figure 6.5b).

BPON is still essentially a synchronous approach. However, replacing the TDMA protocol by an ATM version leads to a form that is more suitable for integration into the ATM B-ISDN [3, 4]. This is an attractive method for the early introduction of ATM to small-business and residential customers. A 155 Mbit/s or 622 Mbit/s cell stream is broadcast from the exchange to the customer terminals which select the cells destined for them. In the reverse direction, a negotiation protocol is invoked that gives each terminal a chance to transmit its cells, which are timed so that a single cell stream is formed in the upstream direction. The ATM protocol has the advantage that, although the total available bandwidth is still shared between customers, the allocation need not be fixed; greater amounts of bandwidth could be given to those customers requiring it at a particular time. This would allow wider bandwidth interactive services to be used. However, such a strategy could lead to blocking if more customers than expected require the larger bandwidth at any one time.

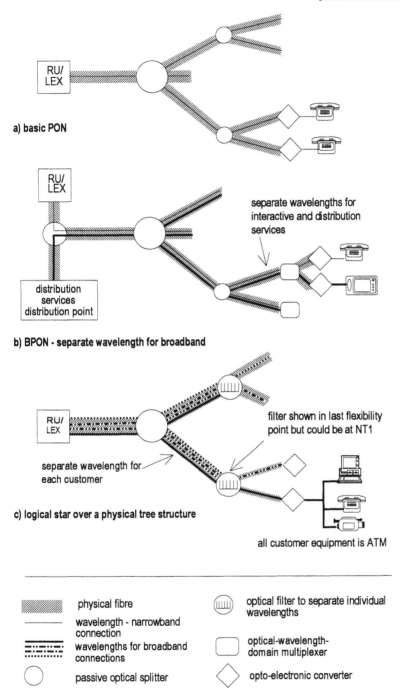

a) basic PON

b) BPON - separate wavelength for broadband

separate wavelengths for
interactive and distribution
services

distribution
services
distribution point

c) logical star over a physical tree structure

separate wavelength for
each customer

filter shown in last flexibility
point but could be at NT1

all customer equipment is ATM

▨▨▨	physical fibre	⊞	optical filter to separate individual wavelengths
——	wavelength - narrowband connection		
▄▄▄	wavelengths for broadband connections	▢	optical-wavelength-domain multiplexer
◯	passive optical splitter	◇	opto-electronic converter

Figure 6.5 PON, evolving through BPON to a wavelength (logical) star

Advantages of passive optical networks:

- Fibre is installed to the home at the outset, so enabling easy migration to non-shared techniques.
- PONs are also suitable as an access method for small businesses, so one technique can be used in a mixed area of residential/small-business customers.
- Use of a tree structure and passive splitters reduces the cost.
- Distribution services can be carried easily on a separate wavelength.
- It could be the final stage if improvements in codec design reduce the bandwidth required for services so that the shared dynamically-allocated bandwidth is sufficient.
- It could be an evolutionary step on the way to a logical-star network carried over a physical tree.
- The possibility of dynamic allocation of bandwidth using an ATM protocol would allow customers to be allocated larger amounts of bandwidth for interactive services than under a static allocation scheme, such as using conventional TDMA techniques.
- Because PON is expected to be competitive with existing CATV networks, satellite television and video tapes, it will encourage the installation of fibre access networks; once these networks are in place the new services could be added relatively quickly making B-ISDN a possibility for the residential customer.

Disadvantages of PONs:

- The bandwidth available is shared between customers: with a reasonable splitting ratio, the bandwidth available to any one customer is fairly low and could be insufficient for some services, such as video retrieval.
- Even with an ATM protocol it is unlikely that shared bandwidth would be sufficient if services requiring reasonably large amounts of bandwidth (such as video library) became widespread.

6.5.3.   Direct fibre / logical star

Obviously, the ideal technical solution to the access network would be to have a star-connected fibre network from each customer to a remote unit or local exchange. This would give each customer maximum bandwidth and allow access to all services. The drawback is the cost. However, improvements in optical technology may well lead to this being a viable solution. This approach would be a logical star network carried over a physical tree structure (Figure 6.5c). Such a network could evolve from a shared-bandwidth passive optical network by adding extra wavelengths for those customers needing exclusive bandwidth as that need arose.

Using current technology with wavelength separation of the order of 2 nm, it is expected that around 30 customers could be served with an exclusive wavelength in both the upstream and downstream direction, giving 155 Mbit/s access for interactive services over one fibre tree. In the medium-term future, with improvements in optical technology, this figure would be expected to rise to over 60 customers. Distribution services would continue to be provided over separate wavelengths common to all customers on that tree.

In the upstream direction, the grating filter would be located at the head end; in the downstream direction the filter could be located in the last flexibility point (the distribution cabinet; i.e. at the 'kerb') or in the NT1. There is an obvious desire to have only simple passive devices in the street and from this point of view the NT1 would be a better location; however, access by the network operator is more difficult if the filter is on a customer's premises and one filter per customer is required. There is also the question of security: having the filter at the last flexibility point prevents a customer having anything other than his own signals on his premises. The preferable location would probably be determined by the particular operational features of a real installation.

Advantages:

• It provides full availability of bandwidth and services to all customers and so could be the target access method.
• A logical star can be an evolution from a PON.

Disadvantages:

• It may be too expensive to be used in the early stages of the evolution.
• It depends on improvements in optical technology.

6.5.4.   *Influence of existing CATV systems*

Existing CATV systems are unlikely to be able to evolve technically to provide B-ISDN access to residential customers, since they generally use copper cable. However, the CATV 'habit' may encourage customers to subscribe to the services carried over a passive optical network and thus may help with the introduction of B-ISDN.

Apart from the varying degrees of cable television penetration across Europe, the situation is further complicated by the potential competition between cable and direct broadcast satellite (DBS) technologies. The arrival of these satellites, with their tendency to be less than respectful of national boundaries, has clouded the plans of network operators and service providers regarding the deployment of optical-fibre cable in the local network. Currently, the distribution of video services, particularly higher-definition TV services, is seen as crucial by the network operators in stimulating the demand for broadband services from residential customers.

6.6. High-speed LANs and B-ISDN

Network operators are currently considering a number of high-speed wide-area network techniques (i.e. running at greater than 2 Mbit/s) to meet the perceived immediate need for LAN-to-LAN interconnection. The major services that are being considered are frame relay and MANs as well as ATM. Introduction of higher-speed LANs in private networks and the subsequent demand for higher-speed public interconnection can be seen as a 'stepping stone' towards ATM and B-ISDN which will include such high-speed data services [51]. At present frame relay exists as a service offered by a number of operators, but it is difficult to predict whether MANs based on DQDB will be widespread, because ATM systems are being introduced in the very near future.

6.6.1. Early interconnection of high-speed LANs

If MANs are introduced, it will first be as isolated islands to offer a data service within major metropolitan areas. Subsequently, they will be progressively interconnected through leased lines supported by the existing transmission network in a meshed network. This will also be the initial means of interconnecting high-speed LANs, with the leased lines being perhaps provided by SDH cross-connects (as part of a managed transmission system). An example of a network configuration is represented in Figure 6.6.

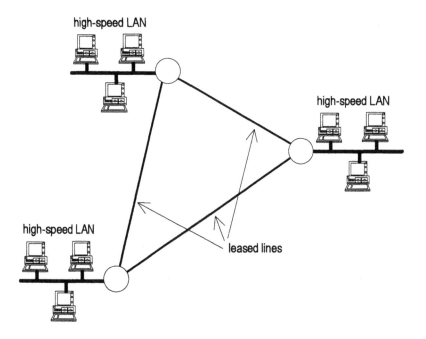

Figure 6.6 First interconnection of MANs with leased lines

6.6.2. *Interconnection through the ATM cross-connect network*

The mesh of leased lines becomes inefficient as the number of networks to be interconnected increases. A first application for the ATM network could be to replace the physical leased lines by VP switching based on ATM cross-connects as shown in Figure 6.7.

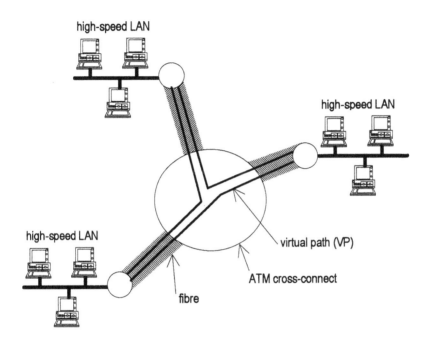

Figure 6.7 Interconnection of MANs by VP cross-connects

Pairs of LANs are connected by a VP through an ATM cross-connect network, with each LAN having a fibre link to one or more cross-connect node supporting all its VPs. Other optical fibre links interconnect the cross-connect nodes themselves and a connection from each LAN to every other LAN is made via pre-allocated, semi-permanent VPs. The cross-connect ATM network provides a virtual leased-lines meshed network. The VPs are set up during initial network configuration and their routeing configuration or bandwidth capacity is re-configured to suit actual traffic demand by the operator of the network from a network-management centre. The conversion between the LAN format and the ATM cell format is handled in equipment at the edges of the network.

Virtual private networks can thus be built over the ATM cross-connect network, providing users with a flexible and rapidly re-configurable network and effective resource and network-management facilities.

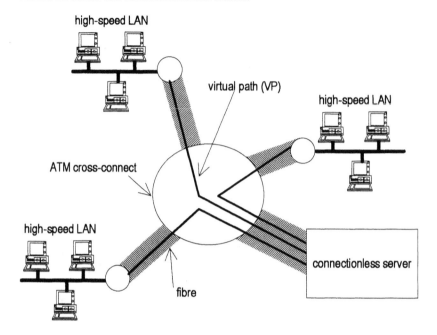

Figure 6.8 Connectionless server in MAN interconnections

6.6.3. *Connectionless data service over the ATM network*

A B-ISDN connectionless data service (CBDS: connectionless broadband data service) allows the transfer of information between users without the need for end-to-end call establishment. Data units can be transferred from a single source to a single destination or from a single source to multiple destinations. Each data unit must include a source address validated by the network. Applications of such a service could be *high-speed file transfer, interconnection of LANs, distributed processing* or *multi-site interactive CAD*.

CBDS will require the introduction of connectionless servers. A 'transit' function will also be needed in the interconnection network to allow an economical connection of any two LANs. In addition to direct virtual paths between some LANs, a semi-permanent virtual path will be set in the cross-connect network between each LAN and a connectionless transit node. This will route the data packets to the proper destination LAN according to the destination user address and, if needed, it will perform interworking between networks using different protocols.

The connectionless transit node is likely to be introduced into the ATM cross-connect network as a centralised server node dedicated to connectionless services (Figure 6.8). Being located in the ATM cross-connect network, it will have ATM interfaces. The ATM connectionless server could also offer a connectionless data service to ATM users connected directly to the ATM cross-connect network for those users who require such a service [70].

6.7. Frame relay and B-ISDN

Frame relay networks (running at lower speeds than MANs) are already being installed to offer semi-permanent virtual-circuit connections for the interconnection of LANs. In contrast to the introduction concept envisaged for MANs, a wide-area network will be installed (instead of MAN islands). These networks will have standard user-network interfaces, but they will be based on proprietary network equipment, developed from the private network domain.

The first opportunity for the introduction of ATM to frame relay networks will be within the switches and the inter-node transmission [69]. ATM might offer a higher throughput than existing proprietary networks; therefore, the existing core network used to support frame relay could evolve to an ATM cross-connect network, with frame relay user-network interfaces.

With the introduction of B-ISDN, terminal adapters will be provided on customer premises that will offer frame relay interfaces to existing customer equipment and interworking with existing frame relay networks will be supplied through gateways between the networks.

6.8. Mobility and B-ISDN

New developments in public land mobile networks (PLMNs) offer the prospect of a large growth in mobile services. Their growing significance for generating revenue makes them especially attractive for both network operators and system suppliers, provided that there is sufficient usage to generate sufficient return on investment. This objective can best be met if a set of flexible and widespread services can be provided that offer a high grade-of-service integrity and good reliability.

First generation cellular PLMNs have been very successful throughout Europe. However, because of increasing demand and a requirement for new and better mobile services, the existing first generation cellular PLMNs will not meet future requirements and will eventually run out of capacity. With recent advances in digital technology and cellular radio techniques, it is now possible to implement a second-generation PLMN with improved features and this is leading to the new GSM (global system for mobile communications) standard. These new PLMN techniques offer higher network capacity and a broader spectrum of flexible mobile communication services including inter-country roaming capabilities, leading to a truly pan-European mobile service.

The ATM B-ISDN will appear soon after these new PLMNs, so it is essential that the ATM broadband network is designed at the outset to interwork with them.

The outcome of ATM development will also be beneficial to the development of the fixed network part of the PLMN, since ATM techniques offer some significant advantages such as: network integration capability, capability for growth, network flexibility, ability for dynamically sharing bandwidth, independence of service and low sensitivity to varying service mix and traffic load.

A third generation of PLMNs is being researched for introduction at the turn of the century, the two systems envisaged being the *universal mobile telecommunications system* (UMTS) and *mobile broadband services* (MBS). UMTS will provide a low-cost pocket telephony service anywhere in Europe as well as higher bit-rate services such as data, thus bringing the features of PLMNs into the domestic and business environment. MBS is planned to supply broadband services from 2 Mbit/s up to 155 Mbit/s (e.g. mobile video), though not initially providing full geographical coverage. The UMTS could actually be integrated with the B-ISDN, with the B-ISDN handling all the mobility functions required by the UMTS. This would have a significant impact on the B-ISDN because location finding and updating require significant signalling and database resources. An alternative would be to provide a gateway between the UMTS and the B-ISDN.

6.8.1. Problems introduced by mobility

The main problems for the network introduced by mobility are the management and control problems of:

- User authentication and registration
- Tracking and location
- Routeing
- Handover
- Charging
- Paging (for calls to the user)

ATM does not in itself solve any of these problems, but it does offer some mechanisms, for example the VP concept, that can help lead to simpler solutions. These problems all introduce additional signalling procedures. For the GSM it is expected that the signalling load (traffic and transaction) will be much higher for mobile services than that for non-mobile. The advent of UMTS will also considerably increase the use of mobile services and the corresponding signalling impact is expected to be very significant. Thus the evolution of the control plane is very important, both in terms of functionality and implementation capacity.

Interference to a radio link can cause a degradation in the call quality, or cause the call to be aborted by the network. This might be a significant problem with ATM because of the possibility of header corruption and this is an area that is still being worked on.

The accumulated delay from the radio interface (the GSM coding/framing delay is approximately 30 ms) and the ATM packetisation delay (if there is no easy translation between the two formats) could lead to a poor quality of service for real-time interactive services such as telephony. This is a potential problem with all mobile services and so is likely to affect UMTS as well. The lower bit-rates for speech commonly used in mobile communications make the situation worse. For instance, the delay in assembling 48 octets of speech at 32kbit/s is 12ms. This is an area where further work is still required in establishing the limits of what is acceptable.

6.8.2. How ATM could influence existing mobile networks

Mobile networks normally use radio links between base stations and the mobile units, but fixed links for the rest of the network (linking base stations to mobile exchanges, and mobile exchanges to each other and other networks). It is this fixed part of the mobile network which is likely to see the most significant impact from the introduction of ATM.

Existing mobile networks have tended to be totally separate from telephony networks with interconnection within the mobile network being handled by leased lines. ATM could alter this in several ways.

Private circuits provided on the ATM fixed network (to interconnect switches in the mobile network) could dynamically share bandwidth with the other services using the ATM network. The bandwidth use of these private circuits could change slowly (to cope with gradually changing demand for mobile access within a given area) or on a call-by-call time scale, or even dynamically within a call, for example, if speech from the mobile network was silence-suppressed.

The ATM fixed network and the mobile network could also be integrated. In this case, there would be no pre-allocated circuits between the switches of the mobile network, ATM connections being set-up on demand using fast signalling protocols.

Chapter 7

Interworking

7.1. Interworking with existing networks

Over the past decade, computer networks have proliferated. With traffic on many networks increasing rapidly, network providers are struggling to cope with their own success. The emergence of new and faster networking technology promises yet more challenges to come.

The introduction of ATM in the near future will be completed in more than one phase; in this period, existing networks will be either integrated into the ATM network or will be interconnected with it. Currently the telecommunications infrastructure within European countries usually contains different networks to serve different purposes:

- Public switched telephone network (PSTN)
- Telex
- Leased lines
- Circuit-switched data network (CSDN)
- Packet-switched data network (PSDN)
- N-ISDN
- TV distribution networks with coaxial cable

The services that the existing networks offer to customers (the range varies depending on the country) will be merged with the new services that the ATM network will introduce. Depending on the introduction approach of the ATM network a proportion of equipment and plant used for these networks will be upgraded, but the greatest proportion will remain as it is (especially in the early phases). There is thus a need for interworking.

In some countries, the older networks, such as PSTN, will have been subsumed by N-ISDN before substantial interworking with B-ISDN is required. In others, the integration with N-ISDN will be slower and may well be overtaken by the introduction of B-ISDN; in these countries there will be a need for direct interworking between B-ISDN and the older networks.

The CCITT has set down a series of recommendations for interworking in its I.500 series. The term interworking is used to express interactions between networks, end systems or parts thereof, with the aim of providing end-to-end communication. Interworking functions may be implemented in the broadband network, in the customer's equipment, in other types of networks, or in some combination of these. The three aspects of interworking that have to be considered are *network interworking, service interworking* and *numbering and addressing*. The first two are treated separately below and *numbering and addressing* is discussed in a later chapter.

7.1.1. *Background to interworking*

Before considering any details of likely interworking, it is important to consider the background and to state certain assumptions.

The first assumption is that narrowband ISDN will be in operation prior to the emergence of ATM, at least in developed countries. Interworking between an ATM network and the narrowband network for voice telephony can then be assumed to be between the B-ISDN and the narrowband ISDN.

It is also assumed that adequate interworking exists between the narrowband ISDN and packet networks and that, in some cases, narrowband ISDN may have incorporated the functions of the original packet switched network. It is likely that new users of packet services will have been connected via the N-ISDN.

Small business LANs that produce relatively low traffic volumes can have a dedicated connection to the ATM network exchange via interworking units which we have termed *LAN adapters*. The need for interconnecting many LANs in large traffic areas can be solved by a backbone MAN that will play the role of a high-speed high-performance public network, collecting large traffic volumes. Such MANs will be interconnected at first with leased lines or digital cross-connects; later they will be interfaced by the switched ATM network.

Where links between customer equipment and an ATM network are considered, the assumption is that a standard B-ISDN interface with ATM capabilities will be provided by the network. If the customer's equipment uses STM operation, then that equipment must incorporate the necessary adaptation.

7.1.2. *Network interworking*

This is the kind of interworking needed when a connection traverses more than one network. It must deal with the following considerations:

- A unique address must be specified to identify an end system. The individual addresses of each end system in a network must be mapped to the appropriate gateway addresses that connect the different networks to the ATM one
- Maintaining the independence of the end-system's addresses from the different network types
- The selection of the path to reach each different destination
- The problems involved with traversing a range of networks when going from source to destination (e.g. PSTN→N-ISDN→ATM)
- The protocol used for accessing each different network on the path
- The packet sizes allowed by each different network

- The specification of a set of quality-of-service characteristics (e.g. delay and cell delay variation, bit-error ratio, cell-insertion ratio) that are appropriate for the service and suitable for all networks on the path

The most important gateway functions that must be implemented are:

- The conversion of information from ATM cells to the appropriate network data units and *vice versa* (user plane functions performed basically by the AAL)
- Signalling conversions (control plane functions)

General guidance on interworking between signalling systems, can be found in Reference 41. Further information on interworking functions of the user plane and the control plane is contained in Reference 39.

7.1.3.   Service interworking

Service interworking (which is really a matter of compatibility between higher layers) is performed whenever there is a conversion of one service to another. For example, when there is a connection between a video-telephone and an ordinary telephone, the video information must be left out and the video-telephony equipment will work as a plain telephone. Another example is the conversion of telex to fax (and such a service already exists) where the interworking functions required will probably be more complex than those of the previous case.

Some services (e.g. standard quality telephony) will exist on both broadband and narrowband networks. In this case, service interworking is not required, only network interworking.

The question raised by service interworking is where in the network will the interworking functions be implemented: in the core network, in the customer's equipment, in the terminal equipment or in a combination of these? In practice, the interworking between one pair of peer layers may occur in a different location from that between another pair.

Signalling interworking is, of course, also important, but signalling is considered as a service so the points discussed above apply to signalling as well.

7.1.4.   Location of interworking functions

In principle, a gateway between two networks can be located at any reference point. The closer the point is located to the terminal, the greater the number of gateways needed and the lower the efficiency of each one of them will be. On the other hand, the amount of traffic per gateway will be lower.

In the customer's network, terminal equipment will be connected via terminal adapters, and small LANs will be connected via 'LAN adapters'. However, gateways (interworking units) will be used for the connections between the ATM network and public MANs, N-ISDN networks and other non-ISDN networks. This was also mentioned in Chapter 4.

As shown in Figure 7.1, public MANs will be connected to the ATM network at exchanges or at remote units; narrowband networks will be connected at exchanges.

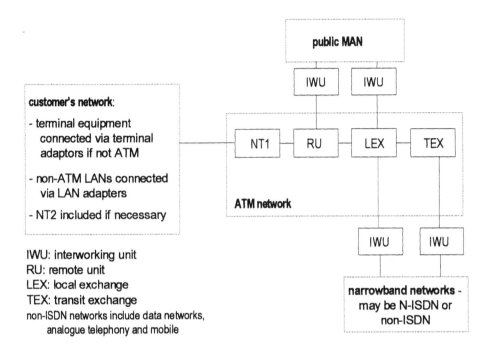

IWU: interworking unit
RU: remote unit
LEX: local exchange
TEX: transit exchange
non-ISDN networks include data networks,
    analogue telephony and mobile

Figure 7.1 Possible location of interworking functions

7.1.5. Gateways handling unlike calls

Gateways handling service interworking must be aware of any limitations in the interworking. Consider the following example covering the call set-up phase between the two users that are shown in Figure 7.2.

User A is connected to LEX_a in the ATM network and has a video-telephony equipment. User B is equipped with an ordinary telephone. Consider the following two cases when user A calls user B:

- User B is connected to LEX_b in the same ATM network
- User B is connected to LEX_b in the PSTN, which is connected to the ATM network via a gateway

In the first case, LEX_b will inform LEX_a that the equipment of user B cannot cope with video-telephony calls. User A's equipment will only be used as a plain telephone and the video service will not be offered.

In the second case, the information about user B's equipment is not required, as the network to which it is connected cannot support video-telephony calls. This fact must be known by the gateway, which will inform the LEX_a so that a plain telephony call will be set-up. If the originating terminal or LEX_a were able to tell from the called number that the destination address was on the narrowband network, it is even possible that the changing of the call attempt to the fallback position could be done by the terminal or at LEX_a (in a similar way to call-barring in existing equipment). This would reduce the signalling traffic across the network.

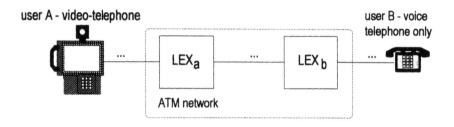

a) video→voice interworking within ATM network

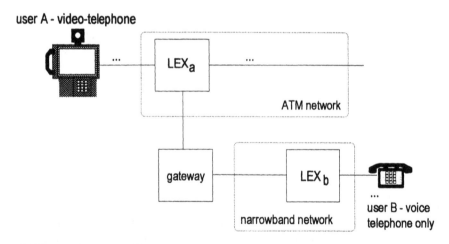

b) video→voice interworking from ATM network to narrowband network

Figure 7.2 Routeing unlike calls

7.1.6.   Impact of ATM on interworking

There will be a problem for interworking with low-bit-rate (16/32 kbit/s) telephony where the delay in filling a complete ATM cell will lead to an unacceptable overall delay. This has not yet been studied within CCITT, but possible solutions could include the use of a special AAL. There will, of course, also be an impact on signalling, because of the need to provide compatibility messages.

Interworking with low-bit-rate telephony will be of most concern when interworking with mobile networks. ATM will be perfectly suitable for carrying multiplexed traffic (for example at 2 Mbit/s) to provide connections between mobile exchanges, but the problem of cell-filling delay will occur when an individual voice call has to be carried between an ATM terminal and a mobile.

Another ATM matter is the question of quality-of-service matching. A user of a virtual channel connection is provided with one of the number of ATM-specific quality-of-service classes supported by the network. There is a different quality-of-service target value for every different service offered. For example, for a file transfer service, the most important parameter is the requirement for very low bit-error and cell-insertion rates regardless of the delay. For a telephone call, the minimum possible delay is required and errors are not so important. Video transmission requires both: the lowest delay and the lowest error rates. During the call set-up phase priorities must be assigned to every service, so that all the required qualities of service will be respected.

An acceptable end-to-end performance must be ensured by all the different networks involved in the service provision. If this target is deemed not to be reachable, the users must be informed so that they can choose to use a service with a lower quality of service, if they so wish.

7.2. Interworking with high-speed data networks

It is expected that high-speed data services will be offered by network operators before the introduction of ATM. Of those under investigation, *frame relay* and *DQDB* (IEEE 802.6) are expected to be the most common. Both approaches offer data services for LAN interconnection and both must interwork with an ATM broadband network [10, 53].

This section highlights some of the major interworking issues that must be addressed to interwork B-ISDN and MAN/frame relay networks.

7.2.1. High-speed LANs and MANs

Interworking between B-ISDN and high-speed LANs based on DQDB will be simplified from the operator's point of view by the adoption by standards bodies of an interface called the inter-MAN ATM interface (IMAI). This is an ATM interface that will allow MANs to be connected together by ATM VPs (Figure 7.3) with the onus for the interworking put on the supplier of the MAN. It will also be required for MANs to interwork with terminal equipment on the ATM network and here interworking units will be required (Figure 7.4).

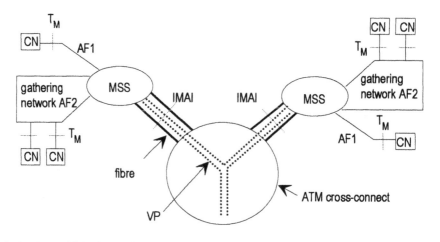

T$_M$: MAN T-interface
AF1 and AF2: MAN access facility
CN: customer network
MSS: MAN switching system
IMAI: inter-MAN ATM interface

Figure 7.3 Interconnection of MANs via ATM cross-connect using IMAI interface

The most likely problems that may arise are caused by the different ways in which isochronous traffic is carried. Within DQDB, voice traffic may be carried by either *pre-allocated* (PA) or *queue arbitrated* (QA) slots. Work in this area is only at a relatively early stage. So far it has been concentrating on using PA slots to carry voice, with these PA slots being set up under network-management control not with call-by-call signalling. The problem for voice interworking is in the way that DQDB and ATM handle voice traffic. Within ATM, a single cell carries traffic from a single source; in DQDB, a single PA slot may carry traffic from a number of sources. In order to interwork, traffic streams will need to be reconstructed, i.e. above the AAL level, which will introduce long delays at the interworking point. The way in which voice is carried might also cause charging problems at the interworking unit. More information about ATM and DQDB interworking is given in Reference 87.

It is also potentially feasible for ATM circuits to carry DQDB traffic, for instance to carry traffic from a DQDB terminal to the main part of the MAN switching system. However, this is now thought to be an unlikely option.

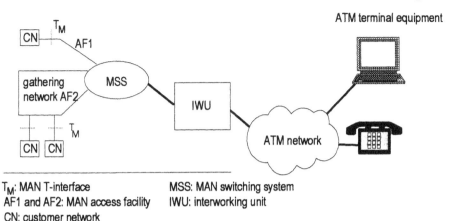

T_M: MAN T-interface   MSS: MAN switching system
AF1 and AF2: MAN access facility   IWU: interworking unit
CN: customer network

Figure 7.4 Interworking of MANs with ATM terminal equipment

7.2.2.   Frame relay

Frame relay is a connection-oriented *access* technique for data transport. Being an access technique, it specifies only a user-network interface and not the networking technique. Frame relay interfaces support a number of virtual circuits, each identified by the digital local connection identifier (DLCI).

In order to support frame relay interworking with ATM, the adaptation equipment must convert from DLCI values to appropriate VPI/VCI values. Signalling information must be converted to the appropriate ATM signalling standards. Frames will be supported by segmentation into AAL 3/4 or possibly by AAL 5.

Figure 7.5 Frame relay services carried over ATM

7.3. Interworking with mobile networks

Voice telephony service is still considered to be the most important service in mobile networks. Digital transmission techniques are employed for this service in the newer mobile networks (such as GSM) and it is envisaged that a bit-rate of 32 kbit/s (50% of the standard 64 kbit/s used for voice) or only 16 kbit/s (13 kbit/s for speech + 3 kbit/s for other) will be used. The reason behind this measure is to maintain high spectrum efficiency at the radio interface, since a lower bit-rate at the radio interface means higher spectrum efficiency. However, using these lower bit-rates introduces additional quantisation distortion and additional transcoder delay which adds to the signal degradation.

If speech signals from GSM telephony services are conveyed via ATM networks, additional delay is introduced because of the cell assembly process and this delay gets longer as the bit-rate is reduced. Interworking of voice services between mobile networks and ATM networks is, therefore, going to introduce additional problems because of this delay. Partially filling the ATM cells may have to be considered as a solution.

Another topic to be considered for interworking is the possible impairment of speech quality caused by lost ATM cells. Again, this may be made worse because of the techniques used in the mobile network to reduce the bit-rate.

Chapter 8

Numbering and Charging

Numbering and charging are two factors that are immediately obvious to the customer and charging in particular can have a significant impact on the take-up of a new service. Historically, in telephone networks, there has been a link between the two: different numbering sequences clearly identify local, national and international destinations, with the consequential differences in tariff. Recently, new, non-geographical codes have been introduced for services such as mobile, premium-rate and free calls thus breaking the obvious link between code and tariff. However, there has been some customer confusion with these new codes, illustrating the importance of numbering to the customer as well as to the network operator.

8.1. Numbering schemes

A proper numbering scheme is necessary to identify unambiguously the correct customer with whom communication is to be established. At present, there are separate numbering schemes (including different country codes) for the main telecommunications services: PSTN, Telex, PSDN. An additional attraction of an integrated services network is commonality of numbering. But this may not be easy considering the present numbering schemes. In the evolutionary stage of the B-ISDN, interworking with existing networks will require the use of the customer numbers for those networks.

The CCITT specifies the ISDN numbering plan in Recommendation E.164. In this recommendation there can be up to 15 digits for a number, as illustrated in Figure 8.1.

The last digit of the subscriber number can be also allocated as a sub-address to select 1 of 8 terminals connected to the ISDN bus. This should not be confused with direct dial-in (DDI) to PBX extensions where a range of public numbers is allocated to the PBX.

The CCITT also allows up to 40 digits (20 octets) for sub-addressing fields as an extension to the addressing capability of the network, but not as part of the numbering plan. The sub-address, when used, will be transferred transparently to the opposite party. Such a supplementary service may be the subject of additional charges and separate registration.

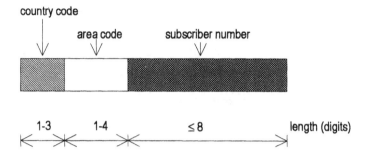

Figure 8.1 Format of E.164 number

8.2. Impact of ATM on existing numbering schemes

In an ATM network, the same principles for numbering and for identification of mobile customers must be used as in the narrowband ISDN. The philosophy of a numbering scheme for a B-ISDN network using ATM should be upwards compatible with that produced for N-ISDN. However, there will be differences because of the capabilities of ATM, and because of the additional broadband services.

Some of the particular points that need considering are:

- It would be helpful to customers if the numbering plan enabled standard teleservices to be readily identified so that customers can associate published numbers with particular services. Broadband ATM will be able to support such a wide range of services that associating numbers with particular services may be difficult to achieve. However, it may be feasible to classify services according to general capabilities, and have a service-class code, either within the national number or as a prefix to the B-ISDN customer number. Then a customer would have a unique B-ISDN number for identification, and one or more possible service prefixes to identify the general service capabilities. The service prefixes could also be used to maintain compatibility with the other network numbering schemes.

  As explained in Chapter 3, broadband services have been classified by the CCITT into the two categories of *interactive* and *distributive* services. These general categories are then divided in service-classes: *conversational, messaging* and *retrieval* for the interactive category; *with* and *without user presentation control* for the distribution category. Within each of these classes, a possible division might be simply into *audio, video* and *data* sub-classes. This three-tier method could form the general capabilities classification for defining service codes. Specific capabilities, e.g. the quality of telephony service, are negotiated as part of the call acceptance procedure.

- The relationship between charging and numbering is much more complex than at present where customers can generally predict the tariff from the number called using the country and area codes. In ATM other factors, such as peak bit-rate, will be used as a basis for charging and this will make it more difficult for customers to estimate the charge when making a call. However, this difficulty will be eased if the numbering scheme also reflects the type of standard service.

- Although not especially a product of ATM, there will be regulatory constraints on the numbering. For instance, within a country with competing administrations, numbers would be seen as a national resource and the allocation would most likely be monitored by a regulatory body.

- Interworking with existing networks may cause numbering problems in the evolutionary stages if, for example, the extra digits of the sub-addressing field of the E.164 number are used.

8.3. Numbering and addressing interworking

When similar networks interwork, it is generally possible to apply a consistent numbering scheme, so that the interworking is transparent to the user. Such a technique has been used for the international telephone network (the concept is inherent in the E.164 numbering recommendation which includes the country code). It is also used for interworking between the land-based and mobile telephony networks: a customer may not realise that a telephone number is actually a mobile and that there is interworking involved in the call. The set-up of the connection to the distant customer is part of the call-control process and numbering interworking is obviously part of this process. Using the transparent numbering scheme across different networks is known as *interworking by call control mapping*.

An alternative technique of interworking exists in which the call set-up procedure uses an address format that explicitly includes the address of the interworking point and the number on the distant network. The call-control information is translated into a format suitable for the other network at the interworking point. This technique is known as *interworking by port access*. The address domains of each different network do not need to be related because the addresses are treated separately.

The most suitable technique for interworking with ATM will depend upon the details of the other network.

8.4. Charging

8.4.1.   General issues

The introduction of ATM to provide a variety of services over a single network introduces a number of problems for charging. Traditional rules for charging based on existing networks (both circuit and packet switched), must be reconsidered so that the charging scheme is seen to be fair for all services carried.

What constitutes a fair charging methodology can be viewed from two opposing camps: that of the customer or the network operator. The former wishes to get the greatest service for the least cost; the latter wishes to get the best return for the investment in new equipment. Obviously, the resulting charging scheme must be a compromise between these positions.

ATM offers the ability to provide a wide variety of services, requiring differing bit-rates and quality of service. This causes a dilemma for the network operator. Charging could be on a flat rate 'per bit' mechanism. This has a number of potential disadvantages. For example, in order to provide a 64 kbit/s telephony service over the ATM network, the per-bit charge must be the same as the existing PSTN. However, this makes high bit-rate services very expensive and might stifle the demand for any new services that require higher bit-rates. Alternatively, the operator may charge differing rates for differing services, making high-bandwidth services much cheaper per bit than lower-bit-rate services. The operator then runs the risk of a customer 'bundling' a number of lower-bit-rate services together to make them simulate a higher-bandwidth service, to take advantage of the apparent discounts available for the higher bandwidth service. A compromise between these positions must once again be found.

The final general issue that must be considered is whether the network operators should charge explicitly for signalling to reflect the network resources that are used to provide signalling, especially as ATM networks will generate much more signalling traffic. However, this might be resisted by customers who have traditionally not paid specifically for it. There might be less resistance if signalling charges were only associated with higher-value services; moreover, the reaction is likely to vary from country to country.

8.4.2.   ATM specific issues

ATM networks offer the ability to support differing services, each of which has a different quality of service. A charging scheme must, therefore, take into account the ability to support such matters as varying levels of burstiness and differing bandwidths. Since an ATM network can also support both multimedia calls and broadcast services, the charging scheme must also be flexible enough to take all these services into consideration. The ways in which resource management and signalling are carried out may also make an impact on the charging scheme. Charging mechanisms must also take into account the effect of signalling and fast reservation protocols, which may allow customers to change their requirements (e.g. bandwidth, quality of service) during a call.

It has been suggested that ATM cross-connects may be installed within the local network. In this case, if there is more than one core network operator (as is the case with the telephony network within the UK), a customer might request one VP to be connected to one core network operator and another VP to be connected to another. This might result in the charges for similar services offered on each VP being different.

Another aspect of charging within an ATM network is the positioning of the charging function, i.e. whether cells are counted as they enter the network, or whether they are counted on arrival. It can be argued that the charging function should be placed at the receiving end in all situations, as cells may be lost in transit, and only those that arrive should be charged for.

8.4.3.   Charging parameters

There are many possible elements on which to base a tariffing strategy for ATM; these have varying significance depending on the particular service being charged for and the characteristics of the particular ATM network. The network should be capable of implementing bandwidth management and charging on each connection of a call. A list of some of the possible charging elements is given below. These elements can be broadly categorised into two groups: those whose value can be defined as a connection is set up and those whose value can change during the course of a call as follows:

Parameters defined as a connection is set up

- Time of day
- Distance
- Bandwidth declared by customer
- Quality-of-service requirements
- Call or connection set-up charge or minimum charge
- Subscription to access classes, service classes or priority

Parameters varying during a connection

- Bandwidth allocated in the network
- Bandwidth used during the call or connection
- Usage (i.e. number of cells or higher-layer packets, transmitted or received)
- Duration of the call
- User-to-user or 'in call' signalling charge

Chapter 9

Routeing Techniques

9.1. Implications for the network structure

The concept of the *virtual path* indicates a distinction between the physical and logical structure of the network. This has an impact on all parts of the network, from the customer access to the long-distance network. The flexibility of the concept is clearly depicted by the example of a mesh logical structure of VPs being implemented on a star physical structure of transmission links (Figure 9.1). The very flexibility of the concept complicates the design task for both physical and logical structures.

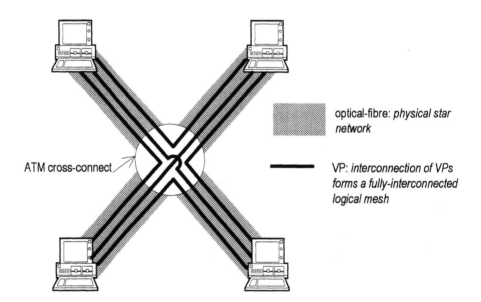

optical-fibre: *physical star network*

ATM cross-connect

VP: *interconnection of VPs forms a fully-interconnected logical mesh*

Figure 9.1 VPs forming a logical mesh over a star-connected physical network

Network structures form the basis of a routeing plan. Different customer groups can be linked by different types of route, choosing between combinations of hierarchies, mesh or star structures, direct, alternative and transit routes. The average and maximum numbers of transmission links between any pair of customers are important parameters in the trade-off between optimum technical performance and economic efficiency.

9.1.1. Routeing of customer access

For a low density of customers, access may be made to a simple gathering point, which has a common physical route to the local exchange. The gathering point may be a junction in a passive optical network, or a multiplexing remote unit. Some cases will involve long distances, and the geography of a country is a key aspect. Natural barriers are expensive to cross, and the use of existing links (roads, railways) is very convenient. The use of optical fibre instead of copper will influence the economics of such situations.

As the density of customers increases, the use of switching remote units can be justified. This reduces the traffic capacity required on the route to the parent local exchange, especially if high-bandwidth distributive services are supplied to the remote unit, for example, by satellite.

For the business community, the use of the VP concept has a significant impact on customer access. Cross-connect or add-drop nodes may be used between the customer and the local exchange for connection to different networks or carriers. This will influence the topology of the local access network.

9.1.2. Routeing between exchanges

The possibilities inherent in the VP concept contribute to the blurring of the logical distinction between local and long distance, and to a decrease in the necessity for a many-layered hierarchy. Low-traffic paths can be routed over a transit network and high-traffic paths implemented as direct virtual path connections.

In the early stages of introduction, the physical local network might be star structured around a cross-connect node rather than around a trunk exchange. With a logical mesh structure between local exchanges, this would remove one level of the call switching hierarchy. As the densities of traffic and local exchanges increase, further cross-connects and direct routes would be justified.

The purpose of the long-distance network is to connect all local-exchange areas and provide access to the international network. If there are few local-exchange areas, they are easily interconnected in a mesh structure. Otherwise there are two options: non-hierarchical, with traffic being routed over an increased number of links; or hierarchical. These options apply both to switching of VCCs and VPCs.

It is likely that a combination will arise: a two level hierarchy of call switching, with the transit level being provided for international calls and routes that do not justify a direct VP between local exchange areas (though they are likely to be transported between local and transit levels by VP); and a non-hierarchical network of cross-connect nodes, supporting a logical network that can vary its structure according to traffic conditions. An illustration of this sort of approach is given in Figure 9.2 where local exchanges are linked to trunk exchanges by VPs; in addition, direct VPs are shown between local exchanges where there is sufficient traffic to justify them. These VPs are routed via cross-connect functions, which in Figure 9.2 are shown as being either in local exchanges or trunk exchanges. However, it would also be possible to have separate cross-connects.

An important feature when considering the routeing plan of any network is *security*. It is essential to minimise any disruption caused by a network fault and this is achieved through diversity in the network. Diversity can be provided at the ATM layer by providing alternative VPs, or at the physical layer using the managed transmission network.

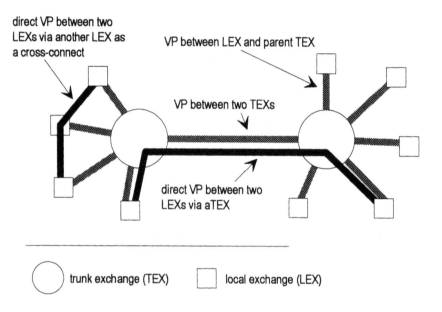

Figure 9.2 Illustration of the use of VPs in the core network

9.1.3.   Distribution of traffic over the network

The expected traffic on the network obviously influences greatly the physical network structure. However, this need not be a restriction when actual traffic turns out to be rather different. The great advantage of the VP concept is that its flexibility enables the network operator to make best use of transmission and switching capacity even though traffic conditions may change. Thus, the design of the physical network structure involves taking into consideration:

• The expected distribution of traffic within and between local exchange areas
• The requirements for alternative routes and network security

- Optimisation of the structure by taking advantage of expected time of day changes in network use.

9.2. Route modification methods

When designing a network, it is important to implement at least one route modification method to cope with variations in traffic load or with system failure. The three techniques applicable to an ATM network are described below.

Alternative routeing

This method provides a call with a choice of two or more routes. It is particularly economical for routes that are expensive to provide: these need to be as fully occupied as possible. In peak traffic conditions, the alternative-routeing algorithm switches calls over second-choice overflow routes if the first-choice high-usage route is full. In an ATM network, these overflow routes may not involve further call switching, because they could be set up as VPCs which just involve a greater number of hops. The ability to employ a number of alternative paths greatly enhances network availability and security, and does not involve reserving transmission bandwidth.

Adaptive routeing

This can be incorporated as an extension to the alternative-routeing method, where the second choice paths are not pre-determined but are established according to the current traffic conditions. Such adaptation to the network state obviously requires more control processing, and may incur greater delays in call establishment. This is traded against the gain in efficient use of transmission resources (at normal network loads).

Dynamic routeing

Dynamic routeing is complementary to the above two, being similar to the adaptive method. However the time scale over which traffic conditions are assessed is different (hours rather than minutes/seconds). In fact, load estimates rather than traffic measurements may be used. The method can exploit, for example, the non-coincidence of busy hours across a large network. If the VP concept is used to the full, dynamic routeing can effectively mean the re-configuring of the logical network structure.

9.3. The routeing plan

A planner has many options when establishing a routeing plan. Decisions are dictated by economics, especially in networks with new routes over long distances. New routes may be limited by the physical infrastructure (such as ducts in the ground). Other plans often depend on the routeing that is available [50]; however, as the network develops, this restriction decreases when enough routes exist to allow the plans to be more sophisticated. For example, the charging plan may depend on the routeing principles being used: charging for telephony has relied on the definition of local calls. But, with the introduction of the VP concept, local routeing will be much more flexible, and the distinction between long-distance and local become more blurred. Technology can also impose restrictions: for example, switching technology may limit the number of links that can be switched. This will have an impact on the possible physical network structure.

The object of the routeing plan is to define all the actual routes which form a network, incorporating both physical and logical structure, and to ensure that they follow a coherent and efficient philosophy. It will include such considerations as:

- The network structure (e.g. hierarchical or fully interconnected mesh)
- The maximum number of transmission links and transit switches allowed in a connection
- The relative merits of direct routes against transit routeing
- The route modification methods that could be used: alternative, adaptive and dynamic routeing
- Implications of switching, signalling, numbering and transmission conversion plans
- The community of interest, and other guidelines justifying direct routes
- Traffic justifying alternative VP routeing and the number of alternatives before a call takes a hierarchical route via a transit exchange
- Dimensioning principles for all route classes
- Customer access and whether remote units are justified and whether remote units should act as concentrators or switches
- Physical and logical structure of the local exchange network

Chapter 10

Traffic Control and Resource Management

The primary role of *traffic control* and *resource management* procedures is to protect the network so that it can achieve the required network performance objectives. The uncertainties of broadband traffic patterns and the complexity of resource management suggest a step-wise approach for defining these parameters and procedures. An initial set of traffic-control and resource-management capabilities is currently defined by the CCITT in Recommendation I.371. Further sets of capabilities may subsequently be defined to achieve increased network efficiency.

In the B-ISDN, congestion is defined as a state of network elements (e.g. switches, concentrators, transmission links) in which the network is not able to meet the negotiated quality-of-service objectives for the connections already established or for any new connection requests, because of traffic overload or control-resource overload. Congestion can be caused by unpredictable fluctuations of traffic flows or by fault conditions within the network. Congestion is to be distinguished from queue saturation, which may happen while still remaining within the negotiated quality of service.

The following functions form a framework for managing and controlling traffic in ATM networks:

- Network resource management
- Connection admission control (CAC)
- Usage parameter control (UPC)
- Network parameter control
- Priority control and selective cell discarding
- Congestion control (selective cell discarding, explicit forward congestion indication)
- Traffic shaping
- Fast resource management

In this book, only the first four functions are considered. Their location in an ATM network is shown in Figure 10.1. The important point to emphasise is that the control of a connection is done on entry to the network and this is explained more fully in Section 10.3.

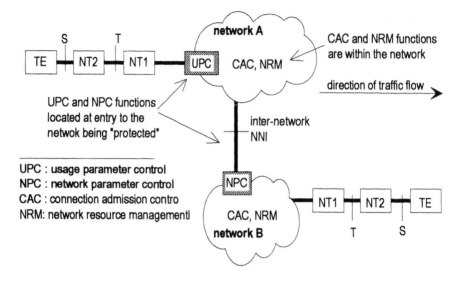

Figure 10.1 Location of traffic-control functions

10.1. Network resource management

Network resource management is responsible for the allocation of network resources in order to separate traffic flows according to different service characteristics, to maintain network performance and to optimise resource utilisation. This function is mainly concerned with the management of virtual paths in order to meet quality-of-service requirements.

Virtual paths are an important component of traffic control and resource management in the B-ISDN. With relation to traffic control, VPCs can be used to:

- Simplify connection admission control
- Implement a form of priority control by segregating traffic types requiring different qualities of service
- Distribute messages efficiently for the operation of traffic-control schemes (for example to indicate congestion in the network by distributing a single message for all VCCs comprising a VPC)
- Aggregate user-to-user services such that the usage parameter control or network parameter control can be applied to the traffic aggregate

VPCs play a key role in network resource management [11]. By reserving capacity on VPCs, the processing required to establish individual VCCs is reduced. Individual VCCs can be established by making simple connection-admission decisions at nodes where VPCs are terminated. Strategies for the reservation of capacity on VPCs will be determined by the trade-off between increased capacity costs and reduced control costs. These strategies are left to an operator's decision.

Where VCCs within a VPC require a range of qualities of service, the VPC is provided with a quality of service suitable for the most demanding VCCs carried. For example, if one of the VCCs within a VPC requires the allocation of the peak cell-rate equal to a significant proportion of the VPC capacity, then assurance of the quality of service of this VCC may require that all other VCCs within this VPC also have an allocation of the peak capacity. The way this will be managed is for further study within CCITT. However, the cell-loss priority bit may be used to distinguish two levels of cell loss on a VPC [24]. The impact of cell-loss priority on the management of the capacity of the VPC is still being studied.

There are three types of application for VPs:

- *User-to-user* application: the VPC extends between a pair of T reference points
- *User-to-network* application: the VPC extends between a T reference point and a network node
- *Network-to-network* application: the VPC extends between network nodes.

With a user-to-user VPC, the network has no knowledge of the quality of service of the VCCs within the VP connection. It is the user's responsibility to determine, in accordance with the network capabilities, the necessary quality of service for the VPC. In the other two cases, the network is aware of the quality of service of the VCCs carried within the VP connection and has to accommodate it.

Statistical multiplexing of VCCs within a VPC where the instantaneous aggregate peak of all VCCs may exceed the virtual path connection capacity, is only possible when all VCCs within the virtual path connection can tolerate the quality of service that results from this statistical multiplexing. The way this should be managed has still not yet been determined.

As a consequence, when statistical multiplexing of VCCs is required by the network operator, VP connections may be used to separate traffic thereby preventing its statistical multiplexing with other types of traffic. This requirement for separation implies that more than one virtual path connection may be necessary between network origination/destination pairs to carry a full range of quality of service between them.

10.2. Connection admission control

Connection admission control (CAC) is defined as the set of actions taken by the network at the call set-up phase (or during call re-negotiation phase) in order to establish whether a virtual channel or virtual path connection can be accepted or rejected. A connection request is accepted only when sufficient resources are available to establish the call through the whole network at its required quality of service and to maintain the agreed quality of service of existing calls. This applies also to re-negotiation of connection parameters within a given call.

In a B-ISDN environment, a call can require more than one connection, for instance for multimedia or multi-party services such as video-telephony/video-conferencing. In this case, CAC procedures should be performed for each VCC or VPC.

In the case of an on-demand service, the connection-establishment procedures will enable the CAC to derive at least the following types of information:

- Source traffic characteristics
- Required quality-of-service class

In the case of a permanent or reserved service (e.g. using a permanent VPC or a permanent VCC), this information is indicated with an appropriate OAM procedure either through an on-line (e.g. signalling) or off-line (e.g. service order) procedure.

Connection admission control makes use of this information to determine:

- Whether the connection can be accepted or not
- The traffic parameters needed by usage parameter control
- The allocation of network resources

Although connection admission control needs information about the traffic characteristics, the methods for characterising traffic have not yet been determined by standards bodies. However, it is expected that traffic characteristics will include measures that describe some or all of the parameters listed below, some of which are mutually dependent.

- Average cell-rate
- Peak cell-rate
- Burstiness
- Peak duration
- Source type (e.g. telephone, videophone)

For a single ATM connection, a user indicates a quality-of-service class from the quality-of-service classes which the network provides. Specific quality-of-service classes are also still being studied.

A variety of connection-admission-control algorithms have been proposed [13, 79]. The aim is to produce an algorithm that is simple (in terms of processing and storage requirements), robust (to guarantee network performance) and efficient (to allow statistical multiplexing gain).

Peak-rate allocation is the simplest approach, but it does not take advantage of the statistical multiplexing available with ATM. Four other approaches are:

- Convolution

- Two-moment allocation
- Linear
- Two level

The current state of knowledge is that there exists a trade-off between complexity and accuracy of the different algorithms. The *convolution* algorithm is relatively accurate, but imposes a high numerical complexity, whereas the *linear* scheme is simple but inaccurate. *Two-moment* allocation leads to satisfactory results only if the peak cell-rate of each connection is small compared to the link capacity.

The *two-level* approach combines the advantages of a fast real-time processing first-level algorithm with more accurate and complex algorithms at the second level. Two strategies have been proposed from work within the European Community RACE programme [85].

The first is characterised by a so-called 'required cell-rate' for all connections in progress. This required cell-rate is updated by the second-level algorithm using the convolution approach and by the first-level algorithm adding/subtracting the peak/mean cell-rate for each connection set-up/release. This is a worst-case assumption.

The second approach is based on a table where the admissible number of connections for different traffic classes is fixed. The second-level connection-admission-control algorithm performs a reallocation of the transmission capacity among different traffic classes based on a more-complex mechanism, e.g. a convolution algorithm or a linear approach.

10.3. Usage parameter control and network parameter control

Usage parameter control (UPC) and network parameter control (NPC) perform similar functions at different interfaces: the UPC function is performed at the user-network interface, whereas the NPC function is performed at the inter-network NNIs (Figure 10.1). Because of the identical nature of the functions being performed, the following sections will consider only UPC, but the same discussion is applicable to NPC.

10.3.1. UPC functions

Usage parameter control is defined as the set of actions taken by the network to monitor and control traffic in terms of conformity with the agreed traffic contract and cell routeing validity at the user access. The main purpose is to protect network resources from malicious, as well as unintentional, misbehaviour that could affect the quality of service of other established connections. It does this by detecting violations of negotiated parameters and taking appropriate actions. This is summarised in Table 10.1.

Connection monitoring encompasses all connections crossing the user-network interface or inter-network interface. Usage parameter control and network parameter control apply to user VCCs and VPCs and to signalling virtual channels. How to monitor meta-signalling channels and the OAM flows has not yet been determined.

The monitoring task for usage parameter control and network parameter control is performed by:

- Checking the validity of VPI and VCI values (i.e. whether or not valid VPI/VCI values have been assigned)
- Monitoring the traffic entering the network from each active VCC and VPC in order to ensure that parameters agreed upon are not violated . This monitoring action is performed at the termination of the first VC link for VCCs and the first VP link for VPCs (Figure 10.2).

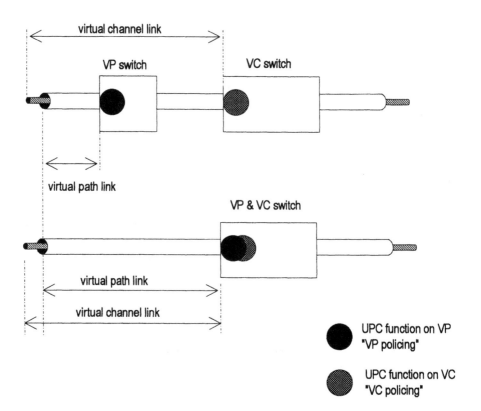

Figure 10.2 Location of UPC functions at end of VP or VC link

The requirement on where the UPC function is to be situated again illustrates the advantages of the VP concept within ATM networks. If a customer has a VP that is established as a user-to-user VP, there is no requirement to perform usage parameter control on the individual VCs, only on the VP as the VC link is not terminated within the network.

If a customer exceeds the agreed parameters, the usage parameter control may take one of the actions listed in Table 10.1. The simplest action is to discard those cells which violate the traffic parameters. In this case the main objective is achieved: a customer will never be able to get more cells into the network than the agreement allows. A more generous (to the customer) alternative is to tag the cells that exceed the pre-negotiated values by setting the CLP bit. If there is no congestion in the network, these cells would be transported to the destination, but if there is congestion, they would be the first to be discarded. This may also be attractive to the network operator as cells carried generate revenue; there may be little point in losing revenue by discarding the cells if they can be carried without harming the network or affecting the quality of service of other connections. A more drastic option is to drop the connection as soon as the contract is violated.

Table 10.1 Options available to UPC for controlling violating cells

discarding cells:	violating cells are dropped
tagging cells:	CLP bit is set so that cells can be discarded later if necessary
dropping connection:	the complete connection is dropped

There is a practical uncertainty in determining the values of the controlled parameters. Hence, in order to have adequate control performance, tolerances of controlling performance parameters need to be defined and, at the time of writing, these are still being studied within the CCITT.

10.3.2. UPC algorithms

A specific control algorithm has not been standardised. However, a number of desirable features of the control algorithm can be identified as follows:

- Capability of detecting any illegal traffic situation
- Selectivity over the range of checked parameters (i.e. the algorithm could determine whether the user behaviour is within an acceptance region)
- Rapid response time to parameter violations
- Simplicity of implementation

Methods to control peak-rate, mean-rate and different load states within several time-scales have been studied extensively [12, 64, 88]. The most common algorithms involve two basic mechanisms:

- The *window* method, which limits the number of cells in a time window
- The *leaky bucket* method, which increments a counter for each cell arrival and decrements this counter periodically

The leaky bucket is generally agreed to achieve the best performance compromise of the mechanisms studied. Its advantages are its simplicity, and flexibility. Figure 10.3 illustrates the principle. A separate control function is required for each VCI/VPI being 'policed'. Every cell arrival increments the counter (the 'bucket'), which is also being emptied at a constant 'leak' rate of at least the mean cell-rate of the connection. If the traffic source generates a burst of cells at a rate higher than the leak rate, the counter fills. Provided that the burst is short, the counter will not fill and no action will be taken against the cell stream. However, if a long burst of higher-rate cells arrives, the counter will overflow and the UPC function will take action against cells in that burst. The tolerance allowed for the connection depends on the size of the bucket and the leak rate.

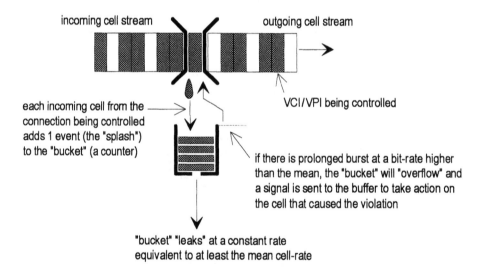

Figure 10.3 Illustration of the principle of 'leaky bucket' control

The major challenge with peak-rate control is the impact of cell-delay variation (CDV), introduced by the access network, on the dimensioning and performance of the different UPC algorithms. One of the conclusions that has been reached within the European Commission RACE programme [85] is that, because of this cell delay variation, the peak rate cannot be controlled adequately by simple mechanisms (particularly those based on the window method). Another parameter, the maximum allowed CDV, must be added to the traffic contract, and this effectively introduces a tolerance to the peak rate control.

However, the consequence of introducing tolerances is to allow traffic with quite different characteristics to conform to the traffic contract. The worst case traffic in this respect is shown in Figure 10.4. The traffic contract is for a CBR connection and, with the CDV allowance, this requires a particular leaky bucket size of, for example, 5 credits. However, this allows a group of 5 cells to pass unhindered at the maximum cell rate of the link! This seriously affects cell loss performance in the network, because such traffic requires significantly more bandwidth to be allocated than that required for the CBR connection.

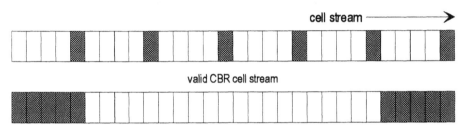

Figure 10.4 Example of the limitations of a leaky bucket UPC with a CDV tolerance allowance

One solution to this problem is the spacer-controller (a combination of a spacing function and a leaky bucket) which enforces a minimum time between cells, corresponding to a particular maximum cell-rate. Spacing is performed only on those cells which conform to the traffic contract, and so this prevents the cell bunching (of the worst case traffic, or caused by variation in cell delay) from entering the network.

Mean-rate control requires a safety margin between the controlled ('policed') and the negotiated cell-rate to cope with the statistical fluctuations of well-behaved traffic streams. This reduces the ability of the UPC function to detect violations of the negotiated mean cell-rate. For example, for a well behaving Poisson source with a mean rate of one cell per time unit, Table 10.2 shows the number of cells which must be accepted within various time intervals in order not to discard more than one out of 10^9 cells.

The shorter the time-scale, the poorer the control of the mean rate because of the large safety margin required. The longer the time-scale, the slower the response to violations of the contract. It is therefore necessary to define a 'virtual mean' over a specified time interval. Several UPC algorithms, for example the moving-window or the leaky bucket, can handle this situation.

More complex UPC algorithms, or a combination of basic UPC mechanisms, have been used to monitor and control different load states of the traffic streams within different time scales. This allows the monitoring of a complete cell-rate distribution (although at the expense of increased control complexity). Examples of such UPC algorithms are the *Gabarit*, the *delta algorithm*, the *multiple leaky bucket*, the *jumping leaky bucket* and an algorithm based on a *nesting principle*.

Table 10.2 Example for mean-rate control

No. of time units:	1	10	100	1000
No. of cells:	12	35	165	1186

A fundamental problem for usage parameter control is that values for traffic parameters have to be anticipated at connection set-up in order to negotiate a UPC contract. This is particularly difficult for services such as data or video applications. These requirements may be less stringent if either a traffic shaping function (see next section), or some traffic classes with pre-defined characteristics, are introduced.

10.4. Traffic shaping

Traffic shaping alters the traffic characteristics (i.e. the inter-arrival times of the cells) of a VCC or a VPC to achieve a desired modification of the traffic characteristics. In its simplest form, this is spacing to smooth out the peaks in cell-rate at the expense of adding in more delay. Figure 10.5 illustrates the principle. Traffic shaping must maintain the cell-sequence integrity of a connection.

Several mechanisms, focusing on different aspects of the traffic streams, are candidates for traffic shaping [85]:

- The cell spacer compensates for cell delay variation and its consequent impact on peak cell-rate by emitting cells at a declared maximum cell-rate (the inverse of the minimum inter-arrival time). This avoids the cell bunching caused by variation in cell delay, and leads to a considerable performance improvement. (Cell bunching reduces the admissible traffic load within the network [7].)
- Another mechanism aims at an optimal scheduling of cells of different connections originating within the same access network. The phase relationship between different periodic cell streams with identical period lengths is adjusted by means of a suitable shaping function.
- A traffic-shaping mechanism that adapts the peak cell-rate at which the next burst will be sent into the network has been used to enforce an effective cell-rate used by the connection admission control. This mechanism is directly related to the linear CAC scheme.

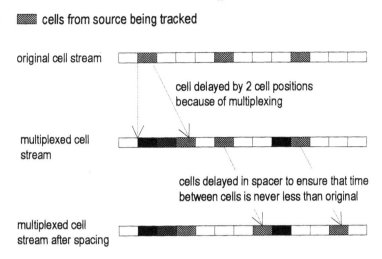

■ cells from source being tracked

original cell stream

cell delayed by 2 cell positions
because of multiplexing

multiplexed cell
stream

cells delayed in spacer to ensure that time
between cells is never less than original

multiplexed cell
stream after spacing

Figure 10.5 Principle of 'spacing'

10.5. Priority control

The user may generate different priority traffic flows by using the *cell loss priority* bit capability. If buffer overflow occurs, network elements are allowed to selectively discard cells of the lower-priority flow, while still meeting the network performance objectives required of both traffic flows.

The simplest scheme, where different routes (virtual paths) are used for the two different priority classes, achieves only a small performance improvement and can be used only if each connection uses just a single cell-loss priority. Most benefit is gained when a connection can use both priority classes.

Two selective cell-discarding schemes have been studied in detail [71, 85]:

- The *push-out* mechanism, where an arriving cell with high priority is able to replace a low-priority cell within the buffer
- The *partial buffer sharing* mechanism, where a part of the buffer is reserved for high-priority cells

The push-out mechanism achieves only a slightly better performance than the partial buffer-sharing mechanism. However, buffer management and implementation are much more complex, because the cell sequence integrity has to be preserved. The partial buffer-sharing mechanism achieves a good compromise between performance and complexity.

Detailed performance studies [71, 85] have shown that the application of cell-loss priorities can lead to a significant improvement of the admissible traffic load.

This allows smaller buffer sizes to be chosen, thus reducing the overall implementation complexity.

10.6. Fast resource management

In response to a user request to send a burst, the network may allocate capacity (e.g. bandwidth, buffer space) for the duration of the burst. Successful operation of *fast resource management* (FRM) schemes requires that this resource allocation be done with minimum delay. When a source requests an increase in peak cell-rate, it has to wait until resources for this increase have been reserved along all the network elements in the ATM connection, with UPC/NPC parameters being adjusted accordingly.

For services with stringent delay requirements, fast resource-management schemes should be implemented in hardware in order to limit the resource-negotiation phase to the propagation delay. For services with relaxed delay requirements, existing signalling procedures as described in [42] could be used.

Fast resource-management schemes introduce a new quality-of-service parameter: *burst blocking probability*. This must be controlled by adequate traffic-control mechanisms.

Different fast resource-management protocols have been proposed for a resource negotiation of either transmission capacity or buffer space. A fast reservation protocol with delayed transmission allocates transmission capacity, in response to a user request, for the duration of one burst. When a source requests an increase of its peak cell-rate, it has to wait until the additional capacity has been reserved on all network elements along the ATM connection before the new peak cell-rate can be used. If this capacity is not available, the new request will be blocked, i.e. the request has to be repeated a short time later. A decrease of the bandwidth requirements is granted immediately, and UPC/NPC parameters are adjusted accordingly. A more detailed description of the protocol and a preliminary performance assessment of the protocol can be found in [85]. Performance results confirm that the protocol is more efficient as a statistical-multiplexing scheme if the typical burst duration is larger than the round-trip delay.

10.7. Congestion control

For low-priority traffic, some adaptive rate-control facilities at the ATM layer or above may be used. Such cell-based reactive techniques are still under investigation.

10.7.1. Selective cell discard

A congested network element may selectively discard cells explicitly identified as belonging to a non-complying ATM connection (i.e. *tagged* by the UPC) or those cells with lower priority (the CLP bit set to 1). This is primarily to protect, as long as possible, high-priority flows, i.e. those with the CLP bit set to 0.

10.7.2. *Explicit forward congestion indication*

A congestion-notification mechanism may be used by end users to assist in recovery during a congested network state. Since the use of this mechanism by the customer equipment is optional, the network operator should not rely on it to control congestion. A network element in a congested state may set an explicit forward-congestion indication in the cell header so that this indication may be examined by the destination terminal equipment. For example, the end user's customer equipment may use this indication to implement protocols that adaptively lower the cell-rate of the connection during congestion. The mechanism should not be limited to data services; variable-bit-rate codecs could change their coding scheme to reduce output rate.

Preliminary studies [59] suggest that a congestion-notification mechanism is of greatest benefit when the duration of congestion is at least an order of magnitude greater than the propagation delay, but it may be of some benefit when the times are of the same order of magnitude. However, when congestion duration is less than the propagation delay, the effect of the mechanism may be harmful both to network operation and user traffic.

Chapter 11

Intelligence in the Network

Network management [92] is the act of controlling, supervising and maintaining communication networks with the ultimate aim of maximising the profit for the network operator, while providing the best possible services and quality of service for the customer. Network management is not only concerned with the automated functions of management, but also the human aspects. It has been described as being about the people, procedures and tools required to manage a telecommunications network effectively at each stage of its life cycle. The term *telecommunications management network* (TMN) describes the overall system used to manage a B-ISDN. It is conceptually separate from the network that it manages, although physically it is likely to share the same media. Associated with network management is the function of *operations and maintenance* (OAM) which is concerned with monitoring and controlling specific entities within a network.

In any network it is also important that the network operator can introduce new or enhanced services readily; it is also desirable that facilities should be provided to allow the user to modify the service/feature to match individual needs. This is likely to be especially true of the emerging broadband ATM network which will support new services, including multimedia applications. It is the purpose of the *intelligent network* to do this without the operator having to load new software into every switch whenever a new service is introduced.

TMN and IN are both aspects of a broader concept: *intelligence in the network*.

11.1. Operations and maintenance principles

Operations and maintenance within the B-ISDN will be required to handle the following functions:

- *Performance monitoring*: Normal functioning of the managed entity is monitored by continuous, or periodic, checking of functions. The result produced is maintenance event information.
- *Defect and failure detection*: Malfunctions, or predicted malfunctions, are detected by continuous, or periodic, checking. As a result, maintenance event information is produced or various alarms are set.

- *System protection*: The effect of failure of a managed entity is minimised by blocking or changeover to other entities. The result is that the failed entity is excluded from operation.
- *Failure or performance information*: Failure information is given to other management entities and, as a result, alarm indications are given to other management planes. Also, responses to a status report request will be given.
- *Fault localisation*: Internal or external test systems interrogate the failed entity if the original failure information is insufficient.

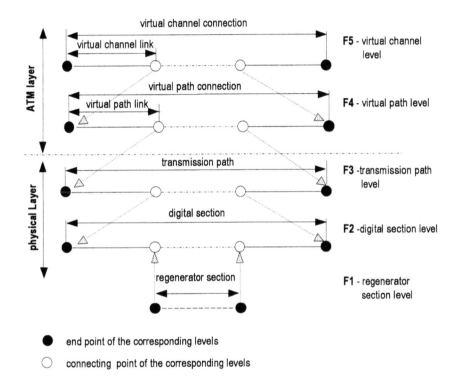

Figure 11.1 Relationship between OAM flow and section levels

OAM functions in the network are performed on five OAM hierarchical levels associated with the ATM layer and the physical layer (Figure 11.1). These operations result in corresponding bi-directional information flows *F1*, *F2*, *F3*, *F4* and *F5* (referred to as *OAM flows*) as shown on the figure. Not all of these levels need to be present as the OAM functions of a missing level are performed at the next higher level. An illustration of the levels is given in Figure 11.2.

11.2. Network management

11.2.1.   OSI functional groups

The OSI functional groupings (Table 11.1) for network-management functions are widely referenced in the literature, for example in [91]. Hence, they will not be considered in abstract here, but only in terms of the implications of having network management within an ATM environment.

Fault management, accounting management and security management will, of course, have to take into account specific ATM factors (such as the methods of charging in accounting management). However, perhaps the most significant impact of ATM [91] will be in the areas of *configuration & name management* and *performance management* and it is only these two areas that are considered here.

Table 11.1 OSI functional groupings for network management functions

functional group	set of functions and tools to:
configuration & name	identify and manage network objects
performance	support planning and improve system performance
fault	locate and repair abnormal operation of network
accounting	support billing for the use of network resources
security	support safety management functions and to protect managed objects

11.2.2.   Configuration & name management

Network management has to know about the configuration of the network and its elements, about possibilities of re-configuring the network for special purposes and about the impact of these reconfigurations on network performance. Each network element must be uniquely identified by its name.

11.2.2.1.   Configuration & name management at the network element layer
Complex network elements like ATM exchanges can also be re-configured for special purposes. The management of these network elements has to know about possible reconfiguration of all the resources of the network element. Specific needs are configuration & name management for CAC, UPC and NPC.

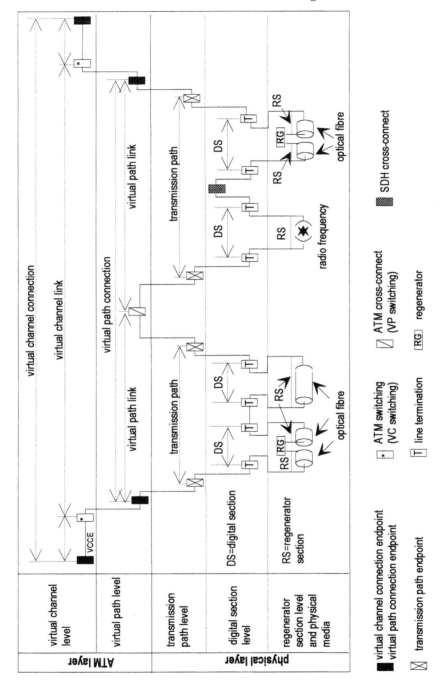

Figure 11.2 Illustration of the different levels within an ATM connection

11.2.2.2. *Configuration & name management at the network layer*

Configuration & name management at the network layer defines, changes, monitors and controls the resources and data required for continuous network operation. It provides facilities for setting, collecting and storing system parameters, changing the system configuration, maintaining a suite of authorised configurations, monitoring the current state and reporting deviations from authorised values.

A reconfiguration can be made either by changing network-element values (configuration & name management at the *network-element* layer) or by choosing another network topology (configuration & name management at the *network* layer). The change of the network topology can be subdivided into *long-term topology tuning* (changes to the static network topology) which is the responsibility of planning and *short-term topology tuning* (i.e. adjustment of bandwidth allocation) which is part of performance management.

11.2.3. *Performance management*

From the viewpoint of the end user, the visible aspects of performance management include response time, rejection rates, availability etc. These may be expressed as agreed levels of network performance.

From the viewpoint of the system manager, performance management is required to provide the information necessary to determine how well the actual network performance compares to the planned network performance, how well user agreements are being met and how the performance of the system can be improved. Performance management must therefore provide the system manager with sufficient tools to provide for:

- *Monitoring*: tracking system activities in order to gather the appropriate data for determining performance
- *Control*: specifically of those elements of the performance-monitoring activity which provide the capability to discriminate between data items which are required by system managers and those currently falling outside a sphere of interest
- *Analysis*: assessment of the results of performance measurements

Evaluation of a system's performance and decisions about the corrective actions may be carried out in different ways. The performance manager or management application process has the responsibility of taking the decisions about the appropriate tuning actions to improve the performance. Two extreme examples are:

- An immediate reaction after a performance degradation has been detected (i.e. after reaching a threshold)

- Collecting comprehensive performance data (statistics) over a long period and applying long-term comprehensive actions resulting in gradual configuration changes

11.2.3.1. *Throughput monitoring function*
Excessive traffic or bottlenecks in the ATM network may be detected by measurement of the throughput of ATM communication channels (measuring either sending throughput or receiving throughput, or both.

11.2.3.2. *Response time monitoring function*
Response time is used to evaluate the quality and effectiveness of an ATM communication network. Response-time thresholds (average and worst-case) are required for detecting performance degradation and generating alarm reports. Response time from the user point of view may be defined as the time difference between issuing a request and reception of the corresponding response.

Response times (such as connection-establishment delay, connection-release delay and transit delay) can be measured for a range of network traffic loads and for a variety of connections.

11.2.3.3. *Statistical analysis functions*
Statistics that are closely related to the network performance (such as utilisation, availability, residual error rate, failure probability) are provided in summary information form by performance-monitoring records or by performance-management logs.

11.2.3.4. *Performance management for CAC*
CAC functions are part of call control and routeing. There is a small part which has to be provided by network management: the provision of the traffic data and the network performance data. This data could be in the form of parameters for connection admission control. The algorithms for the calculation of the parameters and the algorithms used by the CAC to evaluate the parameters have to be provided by traffic management. The execution of CAC is a control-plane function, because after the connection has been accepted a virtual path or virtual channel must be assigned to it.

The modification of routeing tables or routeing information can be done by network management, network performance or traffic management.

11.3. Intelligent networks

Services to be provided by means of Intelligent Network (IN) facilities [28, 31] are likely to be those services that would be difficult to introduce on the basic network for the following reasons:

- Long lead time to deployment
- Large number of exchanges to be upgraded

- Incompatible local exchanges
- Limited switching capacity
- Limited trunk circuit capacity
- Inflexible or dedicated service logic
- Lack of comprehensive database
- Suitable service and network management not available.

Examples of services that are likely to be provided by an IN are:

- Freephone options
- Alternative billing features
- Individual emergency calls
- Universal roaming subscribers (URS)
- Credit card verification
- Virtual private networks
- Wide area CENTREX.

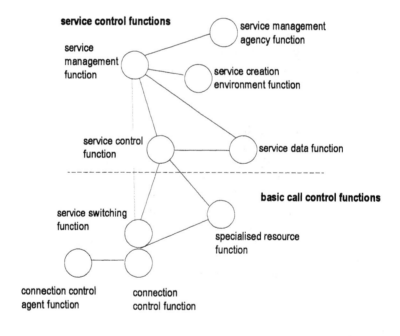

Figure 11.3 Intelligent network functions

It is desirable that the user should be unaware of the actual progress of the connection, whether it is by means of the IN or of the standard network. This implies that a service utilising intelligent-network facilities should aim to complete the connection in no greater time than that expected for a standard network call.

Since a call involving the IN may include interrogation of an external network database this is likely to be a difficult target to achieve.

Intelligent-network type services are activated by a triggered request from the service switching point (SSP) to the service control point (SCP) for a number translation or routeing information. For example, the trigger may be in response to a pre-defined code number normally 'dialled' by the calling party to request a particular service (such as 0800 for 'freephone' in the UK).

There are three distinguishable main processes in the IN:

- Service creation
- Service management
- Service processing

The various functions involved in these processes are shown in Figure 11.3.

11.4. Relationship between TMN and IN

The early use of a fledgling ATM network is expected to be in providing VP cross-connects under network-management control for such applications as interconnection of MANs. Although IN facilities might be useful at that time to allow user control of such things as the reconfiguration of virtual paths, they are unlikely to be essential.

Once call-by-call switching is introduced, IN facilities will be essential. In order to achieve the potential of the network, the operator must be able to introduce new services rapidly in response to customer demand and the IN will be the only practicable means of achieving the required flexibility. At the outset, this must mean that the ATM B-ISDN must not use call-control procedures that are specific to a particular service or a range of services: it must provide a control mechanism that is as universal as its transport mechanism.

The introduction and operation of intelligent-network type services would be controlled via management facilities, although it is probable that the service and network management would be regarded as separately organised, even if they often interrogate and use the same data. The TMN is not only concerned with the management of the network itself and the bearer services it offers; it also manages the more-sophisticated services such as value-added services and those offered by the IN. Therefore, when implementing the IN, developments in the TMN area must be considered in order to achieve the highest possible harmonisation when both concepts evolve. Both the TMN and IN are aspects of a broader concept, that of *intelligence in the network.*

Thus, the IN is to be seen as a large information-processing system, distributed over (and embedded in) the network, consisting of interconnected data processing and storage devices, including network elements equipped with high processing power. It performs its tasks partly under stringent real-time conditions and interacts in a sophisticated way with the human operator: within the IN, the worlds of data processing and telecommunications merge.

IN-based information-handling services comprise two main functional areas:

- The IN service call-handling function involves real-time processing of transient data, i.e. those data items that are only valid for the duration of the respective call. These functions are supported by the communications network in co-operation with service control points which are interrogated during the call for the information to perform the service application.
- The IN service-management functionality deals with semi-permanent data, i.e. those data items defining a service application with its parameters and features. The management of the related functions and objects is proposed to be embedded in the TMN service management layer, while the individual call handling will be the task of the managed network.

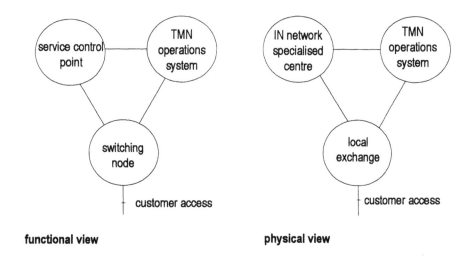

functional view　　　　　　　　　　　　　　**physical view**

Figure 11.4 Advanced services handling triangle

A common set of application functions and application service components will be created to support IN service management and TMN in a consistent way. This allows services to be created by all user groups of TMN/IN (including customers) according to their respective needs and depending on their respective access rights granted by the TMN/IN operator.

The advanced service handling involves a triangle of functional blocks as shown in Figure 11.4.

- Switching and signalling node functions support the call control (i.e. the establishment, rearrangement and release of the individual call). These functions are expected to be in the domain of the organisation providing the basic service and bearer network.
- Network specialised centre (NSC) functions for advanced services cater for the related information processing storage and retrieval necessary to support the

call control. These functions are allocated to the domain of the organisation providing the value-added service.

- IN service management functions are allocated to the TMN service management operation system conceptual level and support the information processing, storage and retrieval required for advanced services management. IN service management functions are assigned to the domain of the organisation that provides the TMN.

It is important to note that an organisation can act in more than one of the roles described above. However, the organisation would have to respect the regulatory constraints imposed, e.g. the equal access rights of value-added service providers to the basic services of the bearer network demanded by open network provision.

Chapter 12

Traffic Engineering

Traffic engineering is about the functional relationships between traffic, resources and performance. It aims to answer questions such as how much traffic needs to be handled, what performance should be maintained, what type and how many resources are required and how they should be organised to handle the traffic?

The CCITT in Recommendation E.700 has structured the traffic engineering tasks in four main sections: traffic modelling, grade of service [62] (performance), dimensioning methods and traffic measurement. Their inter-relationship and dependence on ISDN services, architecture and protocols is illustrated in Figure 12.1.

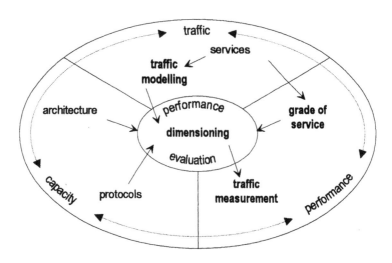

Figure 12.1 ISDN traffic engineering tasks

The focus of Figure 12.1 is the point where architecture, protocols, traffic modelling, and grade of service meet, i.e. in dimensioning methods. For any dimensioning problem there are many alternative solutions. Each solution will attempt to meet the traffic demand and quality-of-service or grade-of-service requirements. However, the network cost is likely to vary according to the solution. The aim of the network planner is to try to optimise the trade-off between network cost, return on investment and quality of service.

In the CCITT GAS handbook 'General network planning' [14], dimensioning is described in the following terms:

> 'To dimension a telephone network is to design the network structure and to determine the amount of equipment required in each part of the network which will satisfy a specific demand with a prescribed quality of service.'

In applying this definition to a B-ISDN based on ATM, the fundamental relationships involving traffic (specific demand), capacity (network structure and equipment) and performance (quality of service) remain, yet the details are different because of the characteristics of the ATM environment [82].

Performance evaluation methods also occupy the same place as dimensioning in Figure 12.1, because they are the necessary precursor to the development of dimensioning methods, and require similar information as input. The emphasis in dimensioning is on determining the capacity, given specific traffic demand and performance targets. Performance engineering is concerned with the same three-way relationship, but is aimed at assessing the feasibility of a particular network design (or, more commonly, an aspect or part of a network) under different traffic conditions. Thus the emphasis is on varying the traffic and measuring the performance for a given capacity (network design). The CCITT uses the term *traffic engineering* as a generic term to cover all areas of traffic, resources (capacity) and (traffic handling) performance. The term *teletraffic engineering* is often used, in order to emphasise the telecommunications nature of the study area.

This chapter treats the traffic engineering tasks (traffic modelling, performance measures, performance evaluation and network dimensioning) of Figure 12.1 in turn. The specific ATM traffic issues associated with traffic control and resource management have been dealt with earlier.

12.1. Traffic modelling

The traditional approach in dimensioning telephone networks has been to model user demand by means of calls. This only requires a few parameters: mean duration and distance (local, national etc.) to describe the call; and call attempt rate to characterise the demand for telephony service from the user.

However, in ISDN traffic engineering, the variety of types of call (depending on the values of the various service attributes) make the modelling of user demand rather more complicated. In the E.700 series of recommendations, the CCITT has developed a systematic approach to describing ISDN calls based on the layered structure for representing ISDN capabilities.

This begins with user demand for information transfer at the user or customer equipment interface. This user demand is transformed by the terminal equipment into call demands for specific ISDN services at the T interface, i.e. traffic offered to layers 1-3 of the ISDN. These transformations may involve functions such as coding, and peer-to-peer and inter-layer protocols (of the higher layers).

A call demand is described by a set of connection characteristics and by a call pattern. The connection characteristics comprise values for those low-layer and general attributes that are relevant for teletraffic engineering, e.g. information transfer mode, information transfer rate, establishment of communication, communication configuration, access protocols (layers 1-3), supplementary services. The call pattern is defined by a set of traffic variables which describe the sequences of events at the T interface and times between these events. These traffic variables are distinguished according to whether they describe events during the call set-up and release phases (call variables), or during the information transfer phase (transaction variables).

This call demand description is the basis for developing user-plane and control-plane traffic models. Consideration of traffic loads from both planes is required for dimensioning, because they may use the same resources (e.g. bandwidth on the access channel). Different traffic flows are distinguished according to a subset of connection characteristics (e.g. information-transfer mode, establishment of communication); then each flow is characterised by a set of variables comprising the call pattern traffic variables and further information based on other connection characteristics (e.g. access protocols).

This systematic approach has been developed from the viewpoint of the narrowband ISDN; ATM considerations imply some changes. The following aspects are affected by ATM:

- Information transfer rate (connection characteristics)
  This will involve a small number of negotiated parameters which will be used as a basis for UPC and CAC. It will make the use of this attribute in distinguishing different types of call demand and traffic flow more complicated unless the negotiated parameters are restricted to a small set of values.

- Information transfer phase events (call pattern)
  call variables: the evolution of signalling capabilities towards a distinction between call control and bearer (connection) control will introduce a layered structure for call-related events.
  transaction variables: ATM introduces events related to the generation of cells, which can be characterised at the burst or cell levels.

- Access protocols (connection characteristics)
  The use of different AAL types will modify the characteristics of the sequence of cell arrival events at the T interface

Information-transfer phase events at the T interface also depend on:

- The type of source information, e.g. voice, video, data
- The transformation in the terminal, e.g. coding, higher-layer protocols
- The presence of physical-layer overhead or OAM cells

12.1.1. *Traffic sources and their behaviour*

Traffic in an ATM network shows behaviour which can be characterised by up to five levels of resolution in time as shown in Table 12.1.

Table 12.1 Characterisation of traffic by resolution in time

Type	Description	Typical duration
calendar	long-term variations	daily, weekly, seasonal
call or connection	delimited by set-up and clear-down events	100-1000 s
dialogue	interaction between users at both ends of an established connection (e.g. silence, A transmitting, both transmitting).	10 s
burst	behaviour of a transmitting user, characterised as a cell-flow rate, over an interval during which that rate is assumed constant	for telephony, the on/off talk-spurts have durations of 0.1 s to a few seconds
cell	behaviour of cell generation at the lowest level, concerned with the time interval between cell arrivals	cell duration: @ 622 Mbit/s: 0.682 µs @ 155 Mbit/s: 2.727 µs

The concepts of user demand and call demand outlined in the systematic approach described earlier help to separate two different types of behaviour associated with traffic sources:

- The characteristic behaviour of a service, in terms of the demands it places on network resources required to support the service
- The use made of the service by telecommunications users, in terms of how often the service is used, and for how long

This distinction in types of traffic behaviour also helps in further distinguishing the primary objectives of dimensioning and performance engineering (Figure 12.2). Dimensioning focuses on the organisation and provision of sufficient equipment in the network to serve the expected use of services made by telecommunications users. It does require knowledge of the service characteristics, but this is in aggregate form and therefore not necessarily to a great level of detail. Performance engineering, however, focuses on the detail of how the network resources are able to support services (i.e. assessing the limits of performance). This requires consideration of the detail of service characteristics, as well as information about typical service mixes (which would be obtained from a study of service use).

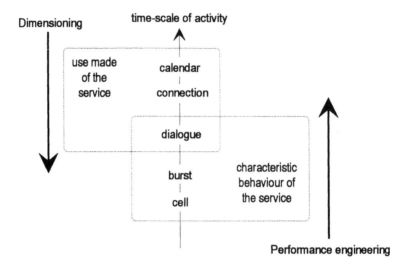

Figure 12.2 Relationship between time-scale, dimensioning and performance evaluation

The CCITT E.700 series Recommendations on traffic-modelling framework provide a detailed means of describing source traffic behaviour, based on service attributes. This primarily addresses the characteristic behaviour of a service (i.e. through connection characteristics and call patterns). However, it can be augmented by the broadband service classification in [26] to include consideration of the application of a service and hence the traffic source behaviour in terms of usage. Such a comprehensive breakdown is a reminder that an application may be supported by more than one type of telecommunications service.

An alternative, very general, classification based on the medium by which information is presented to the user divides services into voice, data and video classes, and this is used in the following discussion on traffic sources.

12.1.1.1. Voice sources

The plain old telephony service (POTS) has traditionally been handled by a fixed-bandwidth channel (whether analogue or digital) which would suggest that a 64 kbit/s channel directly translates to a 167 cell/s fixed-cell-rate source. However, in international circuits, advantage has been taken of the talk-spurt nature of speech (people do not speak for 100% of the time while using the telephone), and the ability of the human ear and brain to compensate for some loss of information, in order to multiplex voice calls statistically and gain better use of expensive international bandwidth by employing silence-suppression techniques. Echo considerations, influenced by cell assembly and disassembly delays, may mean that ATM cells are only partially filled before being transferred over the network. Fixed and variable rate coding schemes, along with silence suppression and partial filling of cells mean that POTS could have a variety of cell stream behaviours. Although it is unlikely, for interworking reasons, that such a variety of methods would coexist just for POTS, other audio services such as hi-fi distribution, mobile telephony, and high-quality telephony may adopt alternative methods.

Enhanced voice services such as conferencing, messaging and IN applications will contribute to different traffic behaviour, particularly in the control plane: they will involve the use of specialised service resources such as conference bridges and voice announcements.

12.1.1.2. Data sources

The communication requirements of present computer systems have been classified [73] according to the intelligence of the machines involved in the communication: time-sharing of dumb terminals on a remote host; transaction-based communication between semi-intelligent workstations and a mainframe; file transfer between remote hosts. Although the volumes of data and the response times of these categories vary, their traffic behaviour is either constant-bit-rate or on/off in nature at bandwidths up to about 2 Mbit/s using connection-oriented communication.

Client-server systems require communications access between local workstations and remote servers for file systems, specialised processing and graphics etc. Communication requirements, particularly for short response times, are such that bandwidths of tens of Mbit/s are needed to a variety of servers in an entirely intermittent manner. This requires connectionless communication. Traffic behaviour tends to be in bursts of constant rate, with the bandwidth being determined by the LAN technology and the current loading of the LAN. It should be noted that satisfactory operation of a LAN depends on the near-instant availability of sufficiently high bandwidth to provide the required short response times.

The control-plane traffic behaviour depends to a great extent on how connectionless services (whether directly supported through CBDS, or indirectly through interconnection of LANs) will have their bandwidth managed (e.g. using fast resource-management, or only negotiating a new fixed bandwidth allocation at infrequent intervals). This choice will also affect the user-plane behaviour of aggregate traffic.

12.1.1.3. Video sources

Video traffic broadly divides into still-picture and full-motion categories [45]. This obvious distinction is useful from the viewpoint of traffic behaviour, because of the difference in timing requirements.

Still picture transmission could find a place in the voice and data categories, e.g. facsimile and computer graphics, respectively. Its communication requirements are the transmission of a certain volume of data within a specified time, thus determining the (constant) bandwidth over that period. Coding, if used, merely reduces the volume of data to be communicated. The development of applications based around still-picture transmission (e.g. tele-shopping, CAD) will have an influence on the control-plane traffic, depending on whether each still picture is managed as a separate connection, or as part of a sequence within one single connection.

Full-motion video introduces the concept of a fixed frame interval for the transmission of a single image as part of a sequence of images. Coding can take advantage of frame-by-frame similarities in the image to reduce the volume of data per frame with the consequence of variable-rate traffic behaviour. The statistical characteristics of video images comprise line, frame and scene correlations and white noise, and they are changed to a greater or lesser extent depending on the type of coding. Frame buffering may be used to remove frame and line correlation. Multi-frame buffering begins to reduce the scene correlations. Spacing cells evenly within a frame (or multi-frame) interval can then reduce the burstiness of the traffic. However, it is at this point that the application of the video is important. Interactive video (video-phone or video-conferencing) will carry an encoded voice component, which is delay sensitive. This limits the amount of buffering that can be used. However, distributive video is not limited in the same way, because the voice component is not interactive.

The application of full-motion video will also influence the traffic behaviour in terms of the service usage. Video-telephony will be used on demand in a similar way to conventional telephony, whereas video-conferencing is likely to be reserved in advance. Video distribution is aimed at the entertainment market, where demand can be adjusted through the mechanism of programme schedules.

12.1.1.4. Multi-media

Multi-media traffic will combine a variety of components from the above three categories, and the traffic can be considered component by component. However, the 'packaging' of such components into particular service offerings will involve specific traffic-control and signalling capabilities. This will have an effect on the control-plane traffic and on the traffic control and resource management, which will have to manage all the components of the multi-media call as a single 'package'.

12.1.1.5. Conclusions

Thus far in this section, the focus has been on the characteristic behaviour of a source in terms of the demands it places on network resources required to support the service. There are still uncertainties about this level of behaviour, e.g. the huge variety of types of scenes for video traffic, or the actual coding methods and protocols that will be used (though this is restricted to a choice from a small number of options).

However, the use made of services by telecommunications users (how often and for how long a service is used), particularly with respect to the penetration of new services, is a significant unknown. This places a requirement on the network planner to dimension a network that can cope with a variety of traffic mixes. Indeed, a significant characteristic of a multi-service network serving both residential and business customer groups will be the variation in the mix of services being used over a 24 hour period. The traditional notion of the 'busy hour', dependent on the business community's demand for telephony, will have to be modified. There may well be other 'busy hours' that have specific characteristics needing special consideration, e.g. early morning news and information services and the evening demand for news, entertainment and tele-shopping.

12.1.2. *Traffic parameters and source models*

It is important to distinguish the different reasons for characterising traffic by the use of traffic parameters or traffic variables.

- *Traffic parameters* (CCITT Recommendation I.371) describe the traffic characteristics of an ATM connection, and are grouped into traffic descriptors for exchanging information between the user and the network. They are specified at the ATM SAP or at some point at the ATM layer where a reference event is defined; such an event takes place *before any ATM multiplexing* and *before any cell delay variation occurs*.

  These traffic parameters are used as the basis for determining the resources needed by a connection, and must be enforceable by UPC and NPC to ensure that the connection does not use more resources than are allocated. However, resource allocation and UPC/NPC must account for the cell delay variation introduced by ATM-layer functions (e.g. cell multiplexing) which may alter the traffic characteristics of ATM connections.

- *Traffic variables* (CCITT Recommendation E.711) describe the sequence of events and the times between events *at the T interface*. They are grouped into call patterns for the modelling of call demand.

Source models are used to represent real traffic at either the ATM service access point, T interface or elsewhere in the network. They must do so in such a way that the traffic characteristics and parameters of the real source and the source model are well matched. The source model can be thought of as a means of encapsulating the relevant traffic parameters or traffic variables in a form that can then be used in analysis or simulation.

Source model parameters are used to define the source model, but these are not necessarily the same as the traffic parameters or traffic variables of the real traffic that the source model is intended to represent. Also, further statistical measures may be used in the parameterisation of the model that would not normally be used as I.371 traffic parameters (e.g. the probability distribution of cell inter-arrival times).

Currently, the following parameters are mentioned as candidates for traffic parameters in CCITT Recommendation I.371: *peak cell-rate, average cell-rate, burstiness, peak duration* and *source type*, although only *peak cell-rate* is currently defined. Obviously, these can (with the exception of *source type*) also be used as E.711 traffic variables or source model parameters.

A wide range of source models is relevant to ATM traffic engineering. A review of the models and their parameters is too detailed for this book. The reader is referred to the literature [9, 47, 54, 63, 68, 72, 86, 93, 94, 96].

There are a number of important criteria [45] for selecting source models: accuracy, generality, suitability for simulation or analysis, model parameterisation method and the choice of, and number of, parameters. These criteria divide into two groups: those that relate the source model with the actual traffic being modelled and those that are concerned with how the source model is to be used.

- *Accuracy*: A source model should approximate a real source in such a way that the relevant statistical characteristics of the traffic are incorporated. These characteristics should be translated into measurable quantities which can be obtained from the real source. For example, if the correlation structure of a source has an important influence on performance then the model should incorporate the main characteristics of this correlation structure.

- *Parameter choice*: An adequate choice of parameters should match well with the real source and with the purposes of the study. For example, a model with parameters successfully used in Reference 67 to approximate average delay does not work well when analysing cell loss. However, an alternative set of parameters for the same model [2] leads to much more accurate results. The number of parameters directly affects the complexity of the model, thus a limited number is preferred.

- *Methods for model parameterisation*: All statistical models need to be parameterised, and the preferred method is unbiased estimation of parameters. This involves equating statistical measures (such as mean, variance and autocovariance of the probability distribution) of the model with those measured from the real source in order to solve for the model's parameters. However, sometimes it is not possible or not applicable to use unbiased estimation; so the model's parameters must be directly measured.

- *Generality*: This is the ability of the model to represent a wide range of traffic behaviour. Once the source model is defined, fitting it to real traffic is achieved by appropriate choice of parameters and their values, without changing the model's definition. How important this criterion is depends on how the model will be used.

- *Suitability for simulation*: Models used in simulation must be statistically stable over the period of interest in the study. If there are N identifiable model states, and the highest level of resolution in time in any state is T seconds, then statistical stability can reasonably be expected over a period of time comparable to NT seconds.

- *Suitability for analysis*: Analytical tractability means that the use of a source model leads to solutions which can be expressed in a form suitable for numerical computation. Using appropriate computation methods, results should be obtainable within reasonable limits on computer resources (memory and processing time) and within required accuracy bounds. In many cases, the complexity of a model may mean that approximate solution methods are necessary. This then introduces another important criterion: the accuracy of the solution method associated with a specific source model.

12.1.3. Traffic mixes

The performance of ATM networks, network elements and network-control functions need to be investigated for a wide range of possible traffic mixes. These mixes can be defined as the number of connections of each type of traffic source, and they are usually assumed to be stationary at the connection level (depending on the actual study). For example, to maintain a constant cell-loss probability, the bandwidth allocated to a variable-bit-rate source depends on the other types of traffic on the link; a larger allocation is needed if the other traffic is predominantly constant-bit-rate [78, 80].

Work within the European Community RACE programme [84] has proposed a number of traffic-mix scenarios:

- Homogeneous traffic, i.e. mixes of identical sources
- Traffic sources with identical burstiness, but different peak rates
- Traffic sources with identical peak rates, but different burstiness
- Two or three variable-bit-rate traffic types, with different burstiness and different peak rates
- Many constant-bit-rate sources, and a few variable-bit-rate sources
- Mixes derived from forecasts

For network dimensioning, the traffic mixes should be as realistic as possible. Focusing on different customer groups (e.g. small, medium and large business, and residential), traffic mixes can be derived from forecasts of busy-hour connection attempts and customer and service penetrations.

12.2. Performance measures

There are several performance measures that are specific to ATM, such as cell-transfer delay and delay variation, cell-loss probability, and cell-insertion rate. Cell delay variation and cell loss, arising from queuing and buffer overflow, respectively, occur because of the characteristics and statistical variation of ATM traffic. They are, therefore, important measures of performance in ATM traffic engineering. Chapter 5 describes how ATM-layer phenomena depend on both ATM-layer and physical-layer performance. Physical-layer effects are not a subject of ATM traffic engineering *per se*. However, because random bit errors can cause cell loss (for example), it is important to be able to distinguish the reasons for cell loss when this phenomenon is measured on a real system.

12.2.1. *Network traffic performance*

The user of an ATM network sees the performance of the network components in an indirect way, observing mainly the *grade-of-service* parameters. The network provider is more interested in the performance and utilisation of the network components. Together these parameters describe the performance of the network. The two important grade-of-service parameters for measuring network traffic performance are the *end-to-end delay* and the *end-to-end loss*.

To study end-to-end performance, reference connections are defined. A reference connection should model a representative or typical connection in an ATM network and the local, national and international reference connections defined for narrowband ISDN can also be used for B-ISDN.

Network traffic parameters can be studied only under nominal traffic load conditions. Therefore, background traffic has to be inserted into the network. For this purpose a traffic-flow analyser [60], which is usually part of a network-planning tool set can be used. With this traffic-flow analyser appropriate load conditions can be determined for all relevant network components.

By using appropriate performance evaluation models, the transit delay times of the switches involved in a reference connection are analysed. From these delay times the end-to-end delay can then be found. From the customer point of view, the mean, variance and distribution of the delay time are of major interest. For example, to dimension the receiver buffer for a CBR terminal, the worst-case delay variance has to be taken into account. Cell-loss probability can also be obtained for the switches involved in a reference connection; the results can then be combined for end-to-end cell loss.

12.3. Performance evaluation

Performance evaluation methods fall into three categories: measurement, simulation and mathematical analysis. Measurement methods require real ATM networks being available for experimentation [57]. General queuing networks of almost arbitrary complexity can be investigated using simulation. However, the simulation of very rare events such as cell loss requires such a large number of cells to be considered that the time involved can become prohibitive. Accelerated simulation techniques, such as parallel implementations [46, 76], variance-reduction techniques (e.g. the RESTART technique which is described in Reference 97) and hierarchical modelling (e.g. cell-rate simulation of queuing at the burst time scale [77, 79, 81, 83]) have been developed specifically for ATM traffic studies. Mathematical analysis, using very sophisticated traffic models, has been applied to ATM performance evaluation. However, even simple analysis can provide useful insight into the basic ATM queuing behaviour.

Congestion, leading to cell delay and cell loss, within ATM network nodes is caused by two different effects. The asynchronous arrivals of cells from different connections cause a short-term (cell time scale) congestion, because several cells may arrive almost simultaneously. If the aggregate cell-rate of all connections exceeds the link rate, a long-term (burst time scale) congestion occurs because of the duration of the active states of those connections. The distinction in queuing behaviour is clearly seen in a graph of cell-loss probability against buffer capacity (Figure 12.3). Cell delay variation depends significantly on the cell-time-scale component, whereas the burst-time-scale component is the dominant factor for cell loss [89].

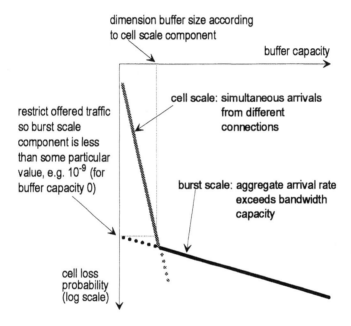

Figure 12.3. Cell and burst time scale components of queuing behaviour

The two different queuing behaviours can be modelled separately, and then the results combined to obtain the overall performance of the statistical multiplexer [75, 90]. Cell-loss probabilities are simply added to approximate the overall cell-loss probability. The cell delay distribution is found from the delay distributions of the separate models by convolution.

12.4. Network dimensioning

The main goal when planning and dimensioning communication networks is to determine the amount of equipment needed to satisfy the traffic demand, while taking several preconditions and restrictions into account. For this purpose, three types of model elements are suggested:

- The logical description of (abstract) traffic within the network
- The switching components in a communication network (for switched services)
- The transmission elements in a long-distance transmission network

The abstract traffic description has to take into account ATM features in detail (e.g. mean and peak cell-rates, traffic source model, traffic mix etc.). Also the models of the switching and cross-connect components in the communication network must be able to describe ATM specific features such as cell loss and virtual paths. Finally, the transmission network (e.g. SDH, or existing PDH, network) has a certain traffic capacity and can only be increased by specific increments.

ATM networks introduce two particular complexities. Unlike the narrowband ISDN, the bandwidth of an ATM connection is not fixed and there is a two-layer switching network (VPCs and VCCs). For network dimensioning, the different bit-rates required are taken into account when considering the demand; the two layers of switching network (Figure 12.4) provide extra flexibility for traffic routes and affect the mapping of the ATM logical network onto the transmission network.

The typical planning procedure can be divided into three steps:

- Describe the end-to-end traffic
- Design and dimension the ATM logical network (switches, cross-connects, links)
- Map the ATM logical network onto a physical transmission network

The main tasks for planning and dimensioning of an ATM-based B-ISDN can be derived from this three-step procedure.

12.4.1. End-to-end traffic

The procedure begins with describing the originated traffic offered to the network using a generalised and well-defined traffic matrix. The traffic matrix includes, for all pairs of source and sink locations, the unidirectional (or bi-directional) offered traffic values.

For ATM, the source traffic for the planning process has to be defined precisely. ATM traffic can be characterised in a simplified form as follows: a group of customers located at node A offer traffic to the network destined for customers at nodes B, C, ..., Z described by a mean bit-rate. This mean value may be calculated from the mean, effective or peak bit-rate related to the modelled service (telephony, data, video). However, this is not just the statistical multiplexing of a particular steady-state mix of services, but an averaging of demand over the whole group of customers in the busy-hour. The variety of traffic mixes in a multi-service broadband network, and the likelihood of a number of different busy hours with specific characteristics, mean that a number of traffic matrices will be needed for the planning and dimensioning process.

An appropriate unit of traffic is required that applies both to the statistical multiplexing of a steady-state service mix, and to the averaging of demand over many customers. Measurement in erlangs does not convey sufficient information, because it needs to be associated with a service bit-rate; hence the *MbitE/s* is used. If the planner knows the average busy-hour demand per customer per service (which can be expressed in erlangs), this can be multiplied by an appropriate bit-rate for each service in the service mix. Totalling these over all services and over all customers in a group gives a MbitE/s value per group of customers. When this is combined with information about the destination of the traffic to all other nodes, the required traffic matrix can be calculated.

Each entry in this traffic matrix (e.g. A_{ij}) includes the originated traffic offered to switch node i and destined for switch node j (i.e. in the call set-up direction). For the opposite direction, a different value may be given, depending on the symmetry and communication configuration of services in the mix. This value is then added to the traffic originated at node j and destined for node i to obtain the value for A_{ji}.

12.4.2. ATM logical network

This step involves two parts: firstly the design and optimisation of a network of ATM 'pipes' grouped into traffic routes between ATM switches and ATM cross-connects; secondly, the dimensioning of the switches and cross-connects to support both access to the network, and the traffic routes between nodes. The design of the traffic routes must also consider the provision of VP cross-connects to provide direct routes between exchanges where there is a high traffic demand.

The design of the logical network begins with the traffic matrix (or matrices) resulting from the first step. Each value A_{ij} in the traffic matrix defines a request for capacity between the two nodes, i and j. Installing a separate traffic route for each A_{ij} gives a fully-meshed logical network with a huge number of small traffic routes; this is normally too expensive.

To optimise this logical network, the traffic of many node pairs is grouped and routed on the same traffic routes by using common transit nodes (these could be either ATM cross-connects or ATM switches). The traditional approach for narrowband networks is to create a hierarchical structure, with each node belonging to a particular level in the hierarchy, and having a parent node in the level above. The longest possible path (the final route) between two nodes is that path which only follows the hierarchical structure. Such a network reduces the number of traffic routes (compared to the fully-meshed network) while increasing the size and number of nodes and the capacity of the remaining traffic routes.

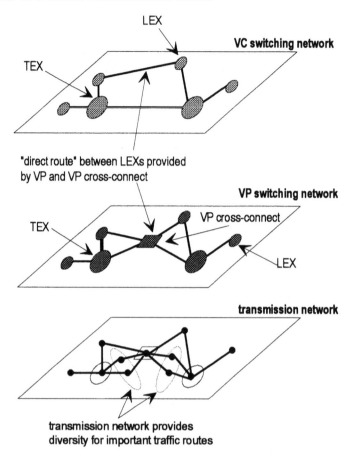

Figure 12.4 Illustration of layering of VP and VC switching and the transmission network

A hierarchical network structure may be optimised by installing additional traffic routes in order to shorten the routeing path (i.e. the number of transit nodes), thus reducing the load on intermediate nodes. One possible normalised cost function for this optimisation is the total cost for switching and transmission. Each additional traffic route increases the transmission cost, but decreases the cost of at least one intermediate node.

The dimensioning of a traffic route and its constituent ATM pipes depends on the required grade of service in the network and the ATM-dependent planning rules. ATM pipes are engineered to a certain maximum utilisation, depending on the number, bit-rate, burstiness and correlation of the connections to be carried by the pipe. Evaluation of this maximum utilisation is done by using methods similar to those used for the connection admission control. Above this level, the ATM network can be treated in a similar manner to STM networks e.g. the dimensioning of a group of ATM pipes (a traffic route) is based on the classical approaches (e.g. the Erlang loss formula) used to determine the capacity of a trunk group.

The dimensioning of the ATM switches and ATM cross-connects comprises two parts: first, calculating the amount of equipment needed at each switching node to allow the offered traffic access to and from the network; second, calculating the amount of equipment needed to connect the traffic routes to the switch or cross-connect to other (destination or transit) switches and cross-connects.

So for each node, the sum of source and sink traffic (i.e. totalling the columns and rows in the traffic matrix) represents the switched traffic of each node (on the customer side) requiring access resources (in terms of ports on the switch). For this, an ATM-based dimensioning rule for ports must be defined. Obviously, this dimensioning is complicated by the type of access (e.g. via a passive optical network, remote unit or direct fibre from the customer).

12.4.3. Physical transmission network

The final step in the network-dimensioning procedure requires calculation of the number and capacity of transmission systems between the trunk transmission stations, and the necessary equipment at those stations. Such equipment includes access ports (for connecting ATM pipes from the nodes onto SDH cross-connects or digital distribution frames) and the multiplexer and cross-connect functions.

The requirements on the transmission network come from the traffic routes between ATM switching and ATM cross-connect nodes. These traffic routes define groups of channels to be installed between associated locations in the network. A separate physical line is not normally planned and installed for each traffic route. The transmission network supplies sufficient transmission capacity for all such traffic routes through a network of nodes and physical line systems created independently of the logical network.

Each traffic route follows a path through the physical network involving intermediate transmission systems and transmission transit nodes. In the most simple case, each traffic route follows only the shortest path through the transmission network. Depending on the definition of 'shortest path' (either distance or hops) the total transmission-system length or the total number of systems in the network can be minimised. In a more realistic situation, each traffic route is split into groups (typically ≥ 2) of ATM pipes, and routed on disjoint paths through the network. This planning strategy takes requirements of high availability into account, supporting network-management facilities e.g. for protection switching. The routeing may also depend on cost characteristics in combination with optimal grouping rules and SDH-ring structures.

The results of this planning step are the numbers of multiplexing and interconnection equipment needed at each trunk transmission station (node) and the number of transmission systems needed between the nodes.

Appendix

Summary of Standards

This appendix briefly outlines the scope of current CCITT recommendations dealing with B-ISDN.

Recommendation I.113: *Vocabulary of terms for broadband aspects of ISDN*
This recommendation consists primarily of those terms and definitions that are considered essential to the understanding and application of the principles of broadband aspects of the integrated services digital network (B-ISDN). They are not exclusive to B-ISDN and are recommended also for application, in so far as they are relevant, to other types of telecommunication network.

Recommendation I.121: *Broadband aspects of ISDN*
This recommendation states the basic principles of B-ISDN and indicates further developments of the ISDN network capabilities in order to support more advanced services and applications.

Recommendation I.150: *B-ISDN ATM functional characteristics*
This recommendation addresses specifically the functions of the ATM layer. This layer is common to all services including signalling and OAM.

Recommendation I.211: *B-ISDN service aspects*
This recommendation should be interpreted as a guideline to the objective of providing detailed recommendations on specific standardised services to be supported by a B-ISDN. The purpose is:

- To provide a classification of such services
- To provide some considerations on the means to describe such services based on the description method as defined in Recommendation I.130
- To give a basis for the definition of the network capabilities required by B-ISDN

Recommendation I.311: *B-ISDN general network aspects*
This recommendation includes networking techniques, application of the virtual path (VP) and virtual channel (VC), VP/VC network element configurations at user access, B-ISDN network-management principles, B-ISDN control and management transport architecture and B-ISDN signalling principles.

Recommendation I.321: *B-ISDN protocol reference model and its application*
This recommendation is based on the ISDN protocol reference model as defined in Recommendation I.320, and wherever not explicitly indicated, it should conform with Recommendation I.320. The purpose of this Recommendation is to take into account the functionalities of B-ISDN, which may eventually be incorporated into Recommendation I.320, as an expansion of the existing ISDN protocol reference model.

Recommendation I.327: *B-ISDN functional architecture*
The general functional architecture model for the ISDN is described in Recommendation I.324. The concepts and associated definitions adopted in Recommendation I.324 also apply to the B-ISDN, i.e. reference configurations, functional groups, reference points.

The objective of this Recommendation is to provide a basic functional architecture of the B-ISDN to complement Recommendation I.324. The model is not intended to require or exclude any specific implementation of the B-ISDN, but to provide a guide for the specification of B-ISDN capabilities.

Recommendation I.361: *B-ISDN ATM-layer specification*
This recommendation specifically addresses the cell structure, the ATM cell coding and the ATM protocol procedures.

Recommendation I.362: *B-ISDN ATM adaptation layer (AAL) functional description*
The scope of this Recommendation is the interaction between different user, control and management requirements on one side and the ATM layer on the other side.

Architecturally the AAL is a layer between the ATM layer and the next higher layer in each of the user plane, the control plane and the management plane. The B-ISDN protocol reference model is given in Recommendation I.321. Examples of services provided by the AAL include: handling of quantisation effect due to cell information field size; handling of transmission errors; handling of the lost and misinserted cell condition; flow control and timing control.

Adaptation layer functions for the control and management planes are not yet defined.

Recommendation I.363: *B-ISDN ATM adaptation layer (AAL) specification*
This recommendation describes the interactions between the AAL and the next higher layer, and the AAL and the ATM layer, as well as AAL peer to peer operations. This recommendation is based on the classification and the AAL functional organisation described in Recommendation I.362.

Different combinations of segmentation and reassembly (SAR) sublayers and convergence (CS) sublayers provide different service access points (SAPs) to the layer above the AAL. In some applications the SAR and/or CS may be empty.

Recommendation I.413: *B-ISDN user-network interface*
This recommendation gives the reference configuration for the B-ISDN user-network interface (UNI) and examples of physical realisations. It describes physical-layer information flows according to the B-ISDN protocol reference model and identifies interface functions. It also addresses OAM issues as they relate to the reference configuration at the user access and to the interface specifications.

Recommendation I.432: *B-ISDN user-network interface: physical-layer specification*
This recommendation defines the physical-layer interface to be applied to the S_B and T_B reference points of the reference configurations of the B-ISDN user-network interface (UNI) at 155 520 kbit/s and 622 080 kbit/s. It addresses separately the physical media and the transmission system used at these interfaces and addresses also the implementation of UNI related OAM functions.

The selection of the physical medium for the interfaces at the S_B and T_B reference points should take into account that optical fibre is agreed as the preferred medium to be used to cable customer equipment. However, in order to accommodate existing cabling of customer equipment, other transmission media (e.g. coaxial cables) should not be precluded. Also, implementations should allow terminal interchangeability.

This recommendation reflects in its structure and content the desire to take care of such early configurations and introduces a degree of freedom when choosing a physical medium at the physical layer. The goal is to have maximum commonality between physical-layer functions at the UNI described in this recommendation and any functions which may be defined in the future at the network node interface (NNI).

Recommendation I.610: *OAM principles of the B-ISDN access*
The scope of this recommendation is to identify the minimum set of functions required to operate and maintain the physical-layer and the ATM-layer aspects of the B-ISDN user network interface (UNI) as well as the individual virtual path (VP) and virtual channel (VC) connections that may be routed through the broadband ISDN. Whenever the term 'customer access' is referred to in this recommendation, it includes the UNI.

The functions of the layers above the ATM layer are not considered for the moment.

Draft Recommendation I.371: *Traffic control and resource management in B-ISDN*
This recommendation describes traffic-control and resource management procedures for the B-ISDN. The main body describes the objectives and framework of traffic control and resource management.

References

1. ANAGNOSTOU, A., CUTHBERT, L.G., LYRATZIS, T. and PITTS, J.M.: 'Economic evaluation of a mature ATM network', IEEE JSAC, Vol. 10, No. 9, pp. 1503-1509, December 1992
2. BAIOCCHI, A., MELAZZI, N.B., LISTANTI, M., ROVERI, A. and WINKLER, R.: 'Loss performance analysis of an ATM multiplexer loaded with high-speed on-off sources', IEEE JSAC, Vol. 9, No. 3, pp. 388-393, April 1991
3. BALLANCE, J.W., ROGERS, P.H. and HALLS, M.F.: 'ATM access through a passive optical network', Electronic Letters, Vol. 26, No. 9, April 1990
4. BALLANCE, J.W. and ADAMS, J.J.: 'Access networks - beyond MANs to solutions compatible with B-ISDN', IEE Conference on Integrated Broadband Services and Networks, London, pp. 58-62, October 1990
5. BLANKERS, P.: 'B-ISDN signalling and intelligent networks', Conference on Integrated Broadband Communication Networks and Services, Copenhagen, April 1993, pp. 22 1.1 -1.11
6. BONATTI, M., CASALI, F. and POPPLE, G. (Editors): 'Integrated broadband communications: views from RACE' (North-Holland, 1991)
7. BOYER, P.E., GUILLEMIN, F.M., SERVEL, M.J. and COUDREUSE, J.-P.: 'Spacing cells protects and enhances utilisation of ATM network links', IEEE Network, pp. 38-49, September 1992
8. BREWSTER, R.L.: 'ISDN Technology' (Chapman and Hall, 1993)
9. BRIEM, U., THEIMER, T.H. and KRÖNER, H.: 'A general discrete-time queuing model: analysis and applications', International Teletraffic Congress ITC-13, Vol. 14, pp. 13-19, June 1991
10. BURAK, M. and GOLDACKER, G.: 'Interconnection of MANs and B-ISDN', First International Symposium on Interworking, Bern, November 1992
11. BURGIN, J. and DORMAN, D.: 'Broadband ISDN resource management: the role of virtual paths', IEEE Communications Magazine, September 1991, pp. 44-48
12. BUTTO, M., CAVALLERO, E. and TONIETTI, A.: 'Effectiveness of the leaky bucket policing mechanism in ATM networks', IEEE JSAC, Vol. 9, No. 3, pp. 335-342, April 1991
13. CASTELLI, P., CAVALLERO, E. and TONIETTI, A.: 'Policing and call admission problems in ATM networks', International Teletraffic Congress ITC-13, Vol. 14, pp. 847-852, June 1991
14. CCITT 'General Network Planning' (ITU, Geneva 1983)
15. CCITT Recommendation E.164 'Numbering plan for the ISDN era'
16. CCITT Recommendation E.700 series on Traffic Engineering
17. CCITT Recommendation G.131 'Stability and echo'
18. CCITT Recommendation G.702 'Digital hierarchy bit rates'

19. CCITT Recommendation G.823 'The control of jitter and wander within digital networks which are based on 2.048 kbit/s hierarchy'
20. CCITT Recommendation G.824 'The control of jitter and wander within digital networks which are based on 1.544 kbit/s hierarchy'
21 CCITT Recommendation I.112 'Vocabulary of terms for ISDNs'
22. CCITT Recommendation I.113 'Vocabulary of terms for broadband aspects of ISDNs'
23 CCITT Recommendation I.121 'Broadband aspects of ISDN'
24. CCITT Recommendation I.150 'B-ISDN ATM functional characteristics'
25. CCITT Recommendation I.210 'Principles of telecommunications services supported by an ISDN and the means to describe them'
26. CCITT Recommendation I.211 'B-ISDN service aspects'
27. CCITT Recommendation I.311 'B-ISDN general network aspects'
28. CCITT Recommendation I.312 (Draft) 'Principles of intelligent network architecture'
29. CCITT Recommendation I.321 'B-ISDN protocol reference model and its application'
30 CCITT Recommendation I.327 'B-ISDN functional architecture'
31. CCITT Recommendation I.328 (Draft) 'Intelligent network architecture'
32. CCITT Recommendation I.35B (Draft) 'B-ISDN performance'
33. CCITT Recommendation I.361 'B-ISDN ATM-layer specification'
34. CCITT Recommendation I.362 'B-ISDN ATM adaptation layer (AAL) functional description'
35. CCITT Recommendation I.363 'B-ISDN ATM adaptation layer (AAL) specification'
36. CCITT Recommendation I.371 'Traffic control and resource management in B-ISDN'
37 CCITT Recommendation I.413 'B-ISDN user-network interface'
38. CCITT Recommendation I.432 'B-ISDN user-network interface - physical-layer specification'
39. CCITT Recommendation I.500 'General structure of the ISDN interworking recommendations'
40. CCITT Recommendation I.610 'OAM principles of the B-ISDN access'
41. CCITT Recommendation Q.601-685 'Interworking of signalling systems'
42. CCITT Recommendation Q.931 'ISDN user-network interface layer 3 specification for basic call control'
43 CCITT Recommendation X.25 'Interface between DTE and DCE for terminals operating in the packet mode and connected to public networks by dedicated circuits'
44. CLARK, M.P.: 'Networks and telecommunications' (Wiley, 1991)
45. COSMAS, J.P. *et al*: 'Source models and applications for video', 2nd RACE workshop on traffic and performance aspects in IBCN, Aveiro, Portugal, January 1992
46. COSMAS, J.P., PHILLIPS, C.I. and MANTHORPE, S.: 'Approaches to simulation of ATM networks and exchanges using distributed processing techniques', IEE International Conference on Integrated Broadband Services & Networks, October 1990

47. COSMAS, J.P. and ODINMA-OKAFOR, A: 'Characterisation of variable rate video codecs in ATM to a geometrically modulated deterministic process model', International Teletraffic Congress ITC-13, Vol. 14, pp. 773-780, June 1991

48. CUTHBERT, L.G. and TORREMANS, E.S. : 'Broadband ATM local networks: evolution and distribution.', IEE International Conference on Integrated Broadband Services & Networks, November 1990, pp.17-22

49. CUTHBERT, L.G.: 'Introduction strategies for ATM ', IEE Colloquium on Fast Packet Switching, January 1991

50. CUTHBERT, L.G.: 'Issues in the planning of broadband ATM networks ', Third IEE Conference on Telecommunications, March 1991

51. CUTHBERT, L.G. and PITTS, J.M.: 'Gateway to broadband: ATM, interworking and evolution', First International Symposium on Interworking, Bern, November 1992

52. CUTHBERT, L.G., HENDERSON, N.G., SAPANEL, J.C., PITTS, J.M., CARETTE, D., KELLER, B, LYRATZIS, T., ANAGNOSTOU, M. and MAIO, J.: 'Introduction strategies for ATM broadband networks in a European context ', Electronics and Communication Engineering Journal, Vol. 4 No. 6, December 1992

53. CUTHBERT, L.G., PITTS, J.M. and SARA, L.: 'MANs and ATM: evolution, interconnection and interworking', SPIE/EOS Conference on Local and Metropolitan Area Networks, Berlin, April 1993

54. DAIGLE, J.N. and LANGFORD, J.D.: 'Models for analysis of packet voice communication systems', IEEE JSAC, Vol. SAC-4, No. 6, pp. 847-855, September 1986

55. DE PRYCKER, M.: 'Asynchronous transfer mode: solution for broadband ISDN' (Ellis Horwood, 1991)

56. DE PRYCKER, M.: 'Introducing ATM for the business and residential customer', 4th IEE Conference on Telecommunications, Manchester, April 1993

57. DE SCHOENMACHER, D. and VERBEECK, P.: 'RATT: a glimpse of a broadband future', Electronics and Communication Engineering Journal, Vol. 4, No. 4, pp. 225-234, August 1992

58. FLOOD, J.E. and COCHRANE, P. (Editors): 'Transmission systems' (Peter Peregrinus Ltd. 1991)

59. GILBERT, H., ABOUL-MAGD, O. and PHUNG, V.: 'Developing a cohesive traffic management strategy for ATM networks', IEEE Communications magazine, pp. 36-45, October 1991

60. GÖTZ, M.: 'TFA_RATT: a traffic flow analysis for the RACE R1022 ATM technology testbed', 2nd RACE workshop on traffic and performance aspects in IBCN, Aveiro, January 1992

61. GRIFFITHS, J.M.: 'ISDN explained', 2nd Edn. (Wiley, 1992)

62. GRILLO, D., LEWIS, A., PANDYA, R. and VILLÉN-ALTAMIRANO, M.: 'CCITT E.700 Recommendation Series - A framework for traffic engineering of ISDN', IEEE JSAC, Vol. 9, No. 2, pp. 135-141, February 1991

63. GRÜNENFELDER, R., COSMAS, J.P., MANTHORPE, S. and ODINMA-OKAFOR, A.: 'Characterisation of video codecs as autoregressive moving average processes and related queuing system performance', IEEE JSAC, Vol. 9, No. 3, pp. 284-293, April 1991

64. GUILLEMIN, F. and DUPUIS, A.: 'A basic requirement for the policing function in ATM networks', Computer Networks and ISDN Systems, Vol. 24, pp. 311-320, 1992

65. HALSALL, F.: 'Data communications, computer networks and open systems' (Addison-Wesley, 1992)

66. HÄNDEL R. and HUBER, M.N.: 'Integrated broadband networks' (Addison-Wesley, 1991)

67. HEFFES, H. and LUCANTONI, D.M.: 'A Markov modulated characterisation of packetised voice and data traffic and related statistical multiplexer performance', IEEE JSAC, Vol. 4, No. 6, pp. 856-867, September 1986

68. HÜBNER, F. and TRAN-GIA, P.: 'Quasi-stationary analysis of a finite capacity asynchronous multiplexer with modulated deterministic input', International Teletraffic Congress ITC-13, Vol. 14, pp. 723-729, June 1991

69. JUVONEN, R.: 'Standardisation of interworking between frame relaying and broadband-ISDN', First International Symposium on Interworking, Bern, November 1992

70. KERKHOF, K. and VAN HALDEREN, A.: 'AAL type 5 to support the broadband connectionless data bearer service', First International Symposium on Interworking, Bern, November 1992

71. KRONER, H., HEBUTERNE, G. and BOYER, P.: 'Priority management in ATM switching', IEEE JSAC, Vol. 9, No. 3, pp. 418-427, April 1991

72. KRONER, H.: 'Statistical multiplexing of sporadic sources - exact and approximate performance analysis', ITC-13, Vol. 14, 1991, pp. 787-793

73. LIDINSKY, W.P.: 'Data communications needs', IEEE Networks magazine, March 1990

74. MANTERFIELD, R. J.: 'Common channel signalling' (Peter Peregrinus, 1991)

75. NORROS, I., ROBERTS, J.W., SIMONIAN, A. and VIRTAMO, J.T.: 'The superposition of variable bit rate sources in an ATM multiplexer', IEEE J-SAC, Vol. 9, No. 3, April 1991, pp. 378-387

76. PHILLIPS, C.I. and CUTHBERT, L.G.: 'Concurrent discrete event-driven simulation tools', IEEE J-SAC, Vol. 9, No. 3, April 1991, pp. 477-485

77. PITTS, J.M., SUN, Z., COSMAS, J.P. and SCHARF, E.M.: 'Burst-level teletraffic modelling: Applications in broadband network studies', Third IEE Conference on Telecommunications, March 1991

78. PITTS, J.M. and CUTHBERT, L.G.: 'Multiservice bandwidth allocation in ATM ', IEE Eighth UK Teletraffic Symposium, April 1991

79. PITTS, J.M., SUN, Z. and SCHARF, E.M.: 'A comparison of burst-level and cell-level approaches to the simulation of ATM networks ', International Teletraffic Congress, June 1991

80. PITTS, J.M. and CUTHBERT, L.G.: 'Traffic mixes on broadband ATM link ', Electronics Letters, Vol.27, No.15, pp. 1350-1351 , July 1991

81. PITTS, J.M., SCHORMANS, J.A. and SCHARF, E.M.: 'Burst Level Simulation: A Comparison with Cell Level Simulation and Queuing Analysis.', 9th IEE UK Teletraffic Symposium, April 1992

82. PITTS, J.M., CUTHBERT, L.G.: 'ATM networks dimensioning: an overview', 4th IEE Conference on Telecommunications, Manchester, April 1993

83. PITTS, J.M. and SUN, Z.: 'Burst-level teletraffic modelling and simulation of broadband multi-service networks', 7th IEE UK Teletraffic symposium, April 1990, pp. 7/1-7/5

84 RACE R1022 deliverable D124, 'Final traffic models and applications', R1022 project office, Alcatel Bell, 1992

85 RACE R1022 deliverable D126, 'Final recommendations on connection admission control, usage monitoring and validation of control schemes', R1022 project office, Alcatel Bell, 1992

86. RAMASWAMI, V., RUMSEWICZ, M., WILLINGER, W. and ELIAZOV, T.: 'Comparison of some traffic models for ATM performance studies', International Teletraffic Congress ITC-13, Vol. 14, pp. 7-12, June 1991

87. RAO, S. and GRÜNENFELDER, R.: 'Interworking of DQDB MAN and ATM-based B-ISDN', First International Symposium on Interworking, Bern, November 1992

88. RATHGEB, E.P.: 'Modelling and performance comparison of policing mechanisms for ATM networks', IEEE JSAC, Vol. 9, No. 3, pp. 325-334, April 1991

89. ROBERTS, J.W.: 'Variable bit rate traffic control in B-ISDN', IEEE Communications Magazine, September 1991, pp. 50-56

90. ROBERTS, J.W. and VIRTAMO, J.T.: 'The superposition of periodic cell arrival streams in an ATM multiplexer', IEEE Transactions on Communications, Vol. 39, No. 2, February 1991, pp. 298-303

91. SCHAPELER, G., SCHARF, E.M. and MANTHORPE, S.M.: 'ATD-specific network management functions and TMN architecture', Electronics and Communication Engineering Journal, Vol 4 no 5, Oct-92

92. SMITH, R., MAMDANI, E.H. and CALLAGHAN, J.: 'The management of telecommunications networks' Ellis Horwood Ltd, 1992

93. SRIRAM, K. , WHITT, W.: 'Characterising superposition arrival processes in packet multiplexers for voice and data', IEEE J-SAC, Vol. 4, No. 6, pp. 833-846, September 1986

94. SRIRAM, K. , MCKINNEY, R.S. and Sherif, M.H.: 'Voice packetisation and compression in broadband ATM networks', IEEE J-SAC, Vol. 9, No. 3, April 1991, pp. 294-304

95. SUN, Z., COSMAS, J.P., CUTHBERT, L.G. and PITTS, J.M.: 'Superposition of GMDP traffic sources for ATM networks', Parallel Processing TENCON '90, September 1990

96. TUCKER, R.C.F: 'Accurate method for analysis of a packet speech multiplexer with limited delay', IEEE Transactions on Communications, Vol. 36, No. 4, pp. 479-483, April 1988

97. VILLÉN-ALTAMIRANO, M. and VILLÉN-ALTAMIRANO, J.: 'RESTART: a method for accelerating rare event simulations', ITC-13, Vol. 15 (Queuing, performance and control in ATM), 1991, pp. 71-76

98. WOODRUFF, G.M., KOSITPAIBOON, R.: 'Multimedia traffic management principles for guaranteed ATM network performance', IEEE JSAC, Vol. 8, No. 3, pp. 437-446, April 1990

Index

Printed in the USA
CPSIA information can be obtained
at www.ICGtesting.com
JSHW011510221024
72173JS00005B/1265

9 780852 968154